FIVE BOROUGH FARM

Seeding the Future of Urban Agriculture in New York City

July 2012

A project of the

DESIGN TRUST FOR PUBLIC SPACE

in partnership with

Added Value

TABLE OF CONTENTS

Eagle Street Rooftop Farm, Brooklyn

Garden of Happiness, Bronx

Phoenix Community Garden, Brooklyn

PREFACE: DESIGN TRUST

Five Borough Farm is a project of the Design Trust for Public Space, in partnership with the nonprofit organization Added Value, to strengthen and expand urban agriculture in New York City. Added Value approached the Design Trust not because of our expertise in urban agriculture, but because of our track record in catalyzing change in the city's public realm. From reimagining the High Line and jump-starting the first purpose-built New York City taxi, to creating sustainable design guidelines for the city's buildings, infrastructure, and parks, the Design Trust produces innovative design and policy solutions to complex, citywide problems.

The vast majority of the more than 700 urban agriculture sites in New York City are on publicly owned land—from schoolyards to the grounds of public housing developments, community gardens to parks—creating a decentralized system of diverse, small-scale, community-based public spaces throughout all five boroughs. These sites do much more than provide fresh produce. They offer educational opportunities, create jobs, capture and reuse stormwater, improve local residents' physical and mental well-being, decrease the waste stream through composting, and bring people together as stewards of communal open spaces. New York now has the opportunity to ensure that the growing urban agriculture movement can continue to make our city a more sustainable, productive, and beautiful place to live.

Five Borough Farm: Seeding the Future of Urban Agriculture in New York City provides the most detailed survey ever produced about urban agriculture in New York City. This publication and the companion website, fiveboroughfarm.org, recommend initiatives to connect farmers' and gardeners' grassroots efforts to municipal policy and provide a framework for understanding and measuring how urban agriculture contributes health, social, economic, and ecological benefits to the city. *Five Borough Farm* also demonstrates the effectiveness of the Design Trust's unique approach. We assembled a multidisciplinary team, including experts in food policy, sustainable design and public health evaluation, a graphic designer and photographer, and engaged more than 100 key stakeholders. Based on this outreach and research, we have documented farms and gardens citywide and identified the urban agriculture community's leading priorities, opportunities, and challenges.

One of the Design Trust's core values is that "social justice and environmental sustainability must guide public space design." Throughout our organization's 17-year history, no other project has embodied this value more than *Five Borough Farm*. With the publication of *Five Borough Farm: Seeding the Future of Urban Agriculture in New York City*, the Design Trust continues its dedicated efforts to make New York City a more environmentally sustainable, dynamic, and healthy city for generations to come.

We are most grateful to Added Value and the Design Trust Project Fellows for their commitment to *Five Borough Farm*, and to the Doris Duke Charitable Foundation and the David Rockefeller Fund for their generous support. Finally, we wish to acknowledge all of the farmers, gardeners, support organizations, government officials, and funders who helped guide and shape this project.

Susan Chin
Executive Director, Design Trust for Public Space

PREFACE: ADDED VALUE

For more than a decade, Added Value, a nonprofit organization dedicated to promoting the sustainable development of Brooklyn's Red Hook neighborhood, has been working on the ground in our community, with sister organizations here in New York and colleagues around the country and the world, to build a healthier and more just and sustainable future for us all. Working together with teams of youth, committed volunteers, partner organizations, and government agencies, we have transformed more than four acres of underutilized City and State parkland into vibrant urban farms where young and old can grow and learn as they sow seeds and harvest crops.

Using urban agriculture as our platform, we have developed a Youth Empowerment Program, which has provided long-term engagement for more than 200 teenagers, and established a Food and Farm Based Learning initiative, which works with more than 1,000 students a year from nearby schools. The food grown on our farm supplies 70 families with weekly vegetables through the Red Hook Community-Supported Agriculture program, and is sold alongside regional produce at our farmers market and distributed to six locally owned restaurants. This past year we composted more than 160 tons of food waste, decreasing the carbon footprint of our neighborhood while increasing the productivity of our farm. Together with friends and neighbors we are building a small yet meaningful economic engine while educating a new generation of young leaders.

As our work has grown and expanded, so too has the field of urban agriculture and the food justice movement. More and more people here in New York City and across the nation realize that our broken food system has serious consequences for our individual health, and for the health of our environment and our economy—particularly for people of color and those in poor communities. It has become more and more important for all of us to be able to understand and articulate how urban agriculture can contribute to our society and economy, and the planet on which we all live.

Because we believe in the potential for urban agriculture to have powerful impacts, we teamed up with our friends at thread collective, a Brooklyn-based design firm, and crafted a proposal to work with the Design Trust for Public Space. I am excited about the work that the Design Trust team has done, and grateful for all those who have contributed their time. We look forward to working with our colleagues in the urban agriculture community to implement the metrics framework and policy recommendations in this publication. I am hopeful that *Five Borough Farm: Seeding the Future of Urban Agriculture in New York City* will contribute to the field of urban agriculture on the ground, in our communities, boardrooms, and classrooms.

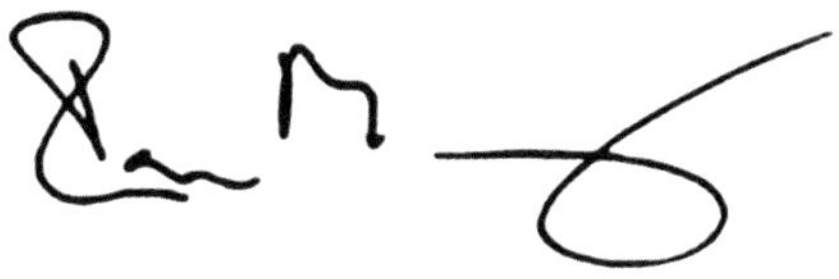

Ian Marvy
Co-Founder and Executive Director, Added Value and Herban Solutions Inc.

EXECUTIVE SUMMARY

Five Borough Farm is a project of the Design Trust for Public Space, in partnership with Added Value, which operates the 3-acre Red Hook Community Farm in Brooklyn. *Five Borough Farm* has three main goals:

1. Document New York City's existing urban agricultural activity through photographs, maps, infographics, and detailed interviews with key stakeholders, and describe the opportunities and challenges facing the city's urban agriculture community.

2. Establish a shared framework and tools to allow users to track urban agricultural activities citywide, and evaluate their social, health, economic, and ecological benefits.

3. Develop policy recommendations that will help make urban agriculture a more permanent part of the city's landscape and governance.

This summary includes key findings and recommendations from the publication's three main chapters – Urban Agriculture in NYC, Metrics, and Policy – and outlines steps to implement these recommendations.

Urban Agriculture in NYC

There are more than 700 farms and gardens citywide that grow food.

This total includes nonprofit and commercial farms, as well as:

- An estimated 390 of 490 community gardens[1]
- 245 of an estimated 600 New York City Housing Authority (NYCHA) gardens[2]
- 117 of an estimated 289 public school gardens[3]

This count does not include sites associated with restaurants or other businesses not solely dedicated to growing food; institutions such as assisted-living facilities or shelters; or individual backyard or rooftop gardens growing food.

The vast majority of these sites are roughly 5,000 square feet or less, with many as small as 2,500 square feet—about the size of three NYC subway cars.

Urban agriculture is about much more than just growing food.

Growing food is often one among many urban agricultural activities directed toward community development goals. These activities include education programs for students, paid internships and youth leadership training for teenagers, nutrition and cooking classes, environmental conservation through composting and rainwater collection, and many others.

Farmers and gardeners face challenges obtaining critical resources.

These resources include physical components such as growing space, soil and compost, and construction materials, as well as nonphysical components such as funding and skilled labor. In interviews conducted for this project, numerous farmers and gardeners said they would like greater involvement in citywide policymaking that affects the availability of compost, land, and other resources.

1 Community Garden Survey 2009/2010, GrowNYC and GreenThumb

2 According to NYCHA records, as of October 2011

3 According to GrowNYC records, as of August 2011

There are race- and class-based disparities that hinder access to information, services, and funding among farmers and gardeners.

Stakeholders interviewed for the project cited race- and class-based disparities in farmers' and gardeners' access to philanthropic dollars, government grants, in-kind assistance, as well as information about these opportunities.

There are few coordinated efforts in NYC to track urban agricultural activities or evaluate their benefits.

There is a lack of basic information about urban agricultural activity in New York City, including a comprehensive count of the number of urban agriculture sites citywide, the number of people who participate in urban agricultural activities, or the amount of food grown in the city's farms and gardens. Many interviewees said they wanted to better understand how urban agriculture contributes to health, social, economic, and ecological benefits, but lacked the time, resources, tools, or technical knowledge to systematically track urban agricultural activities and collect data that would enable them to estimate these impacts.

City agencies generally lack the authority and resources to address urban agriculture.

Without a Mayoral directive or other legally binding mandate, even supportive City agencies have limited capacity to coordinate activity with other agencies, incorporate urban agriculture into existing, potentially complementary plans and programs, or adopt citywide urban agricultural policies and procedures.

Metrics

Numerous studies have demonstrated that urban agricultural activities improve healthy eating, increase physical activity, provide employment opportunities, and build community cohesiveness.[4] However, very little of that work focuses specifically on New York City, or develops measures that can demonstrate benefits at a neighborhood as well as a citywide scale, and ultimately inform the decision-making of elected officials and agency commissioners.

4 See the Metrics chapter for a full set of citations of peer-reviewed studies demonstrating the benefits of urban agriculture.

This chapter establishes a framework to understand how the broad range of activities taking place at the city's farms and gardens can contribute to social, health, economic, and ecological outcomes. Also included are more than forty "indicators," or signs of progress and change, that farmers and gardeners can track. Ultimately, data collected using these indicators could inform citywide urban agriculture policy.

Based on the priorities and capacity of the vast majority of the city's farms and gardens, two types of indicators are recommended:

Indicators to track the scope, scale, and geographic concentration of urban agricultural activities

These indicators can be useful benchmarks for farmers and gardeners to understand the impacts of their programs, and changes over time. Because these indicators can be tracked at any number of sites, the data also can be aggregated to demonstrate the scope, scale, and geographic concentration of urban agriculture citywide. Examples include:

- Pounds of food produced
- Number of young people participating in a farm/garden program
- Number of people employed by a farm/garden program
- Number and type of activities related to increased biodiversity (beekeeping, native planting program, land restoration)
- Dollar amount ($) and percentage of sales that are from food access programs such as the Supplemental Nutrition Assistance Program (SNAP)

Indicators to track outcomes such as health and behavioral changes

These indicators address more complex questions of impacts on behavior and learning through participant surveys. Developing and administering these surveys, and analyzing the data, requires more time, resources, and technical assistance than do the indicators described above. Examples include:

- Number and percentage of people participating in a health/nutrition program who report eating the federal government's recommended five fruits and vegetables per day
- Number and percentage of people participating in a health/nutrition program who can identify where their food comes from (i.e., origin of food as plant based)
- Number of youth farm/garden participants who graduate from high school
- Farm/garden participants' perceptions of safety (personal safety, crime) in their community

Policy

The Policy chapter includes recommendations to integrate urban agriculture more fully into New York City's physical landscape and government; coordinate actions across multiple agencies; and leverage existing investment and programs to address pressing municipal goals. These recommendations build on existing policy documents, such as PlaNYC, Mayor Michael Bloomberg's citywide sustainability plan, and FoodWorks, City Council Speaker Christine Quinn's citywide food policy platform. The recommendations and are organized into four main categories:

1. Formalize government support for urban agriculture

Two critical positions within city government should be strengthened and supported with additional staff and budget to address urban agriculture:

- GreenThumb, a program of the Department of Parks & Recreation, provides a wide range of technical assistance and materials to farms and gardens citywide, and coordinates with other agencies to make city-owned land available for new farms and gardens. The equivalent division within Seattle city government receives roughly the same budget to administer less than one-sixth the number of sites.

- The Food Policy Coordinator is responsible for coordinating city agencies to improve New Yorkers' access to healthy food. This position could be broadened to lead all efforts related to citywide urban agriculture policy.

The Food Policy Coordinator, working with the Department of City Planning, should develop and adopt an urban agriculture plan that:

- Establishes goals and objectives (e.g., increase total number of farms and gardens by a certain percentage, or increase food waste captured and composted at farms and gardens by a certain amount)

- Assesses the economics of urban agriculture, including the potential for food production, costs of expanding production, etc.

- Identifies locations where urban agriculture should be encouraged

- Determines the long-term capital and annual operating budgets, and infrastructure required to adequately support the city's urban agriculture programs

- Creates processes that ensure coordination across city agencies (and between city and state/federal agencies) to support urban agriculture

2. Integrate urban agriculture into City policies and plans

By incorporating urban agriculture in their planning and operations, City agencies could potentially find cost-effective initiatives that would simultaneously achieve their goals, address other pressing citywide issues, and support farmers and gardeners. Two examples are:

- The Department of Environmental Protection (DEP), through its Green Infrastructure Program, has committed $187 million over the next four years to small-scale, low-tech landscape design techniques to absorb stormwater. DEP could prioritize support for urban agriculture projects that also meet the stormwater management objectives of the program, thereby addressing multiple municipal needs simultaneously.

- The Department of Sanitation produces compost from a number of sources citywide, but in 2008 discontinued a program that made this compost available to New Yorkers for free. Sanitation could assess the compost needs of the city's farms and gardens, expand compost production, and reinstate its program to distribute free compost to New Yorkers.

3. Identify innovative opportunities to build urban agriculture into the cityscape

By modifying existing regulations and procedures for developing city-owned land, government agencies have many opportunities to facilitate grassroots efforts and encourage the creation of new farms and gardens. Examples include the following:

- The Department of Housing Preservation and Development (and other agencies) should encourage the incorporation of urban agriculture (and food access, more broadly) into new development projects in their program and design guidelines, and in requests for proposals.

- The Department of Buildings and the Department of City Planning should encourage temporary urban agriculture projects on the more than 600 stalled development sites citywide.[5]

- The Economic Development Corporation should assess the costs, benefits, and potential locations of urban agriculture "food hubs"—locations where farms and gardens could refrigerate, process, and distribute produce.

5 Arrested Development: Breathing New Life Into Stalled Construction Sites, a report by the Office of Manhattan Borough President Scott M. Stringer, p. 1, September 2011

4. Address race- and class-based disparities in New York City's urban agriculture community

Elected officials and government agencies should:

- Increase access to information about available resources, including a list of sources of public funding distributed to farms and gardens each fiscal year
- Support capacity building, including training and assistance with grant-writing, management, and business planning, as well as technical assistance and funding for community organizing and outreach specifically targeting underserved groups
- Facilitate participation in policymaking, including creating a citywide Urban Agriculture Task Force to include representatives from farms, gardens, support organizations, and other key stakeholders, and establishing transparent, citywide procedures for distributing city-owned land and other resources

Finally, this section offers broad suggestions to the philanthropic community to target their investments to support urban agriculture, including expanding grant opportunities for under served communities, exploring a sustainable funding source for farms and gardens, and funding more networking opportunities for farmers and gardeners.

Next Steps

Over the next year, the Design Trust will distribute its findings and recommendations to the public through a series of videos and community workshops. Working with key stakeholders, the Design Trust also will begin to implement key recommendations in this publication, including:

- Providing resources and technical assistance to a representative sample of farmers and gardeners citywide to track common indicators using the *Five Borough Farm* metrics framework

- Making available an online tool on the *Five Borough Farm* website allowing farmers and gardeners to upload data and download reports about their activities

- Establishing a citywide urban agriculture task force for farmers and gardeners, City officials, funders, and support organizations to share information and help shape citywide policy

- Developing criteria to assess the suitability of City-owned land for urban agriculture

Abu Talib, Taqwa Community Garden, Bronx

INTRODUCTION

At Taqwa Community Farm in the Bronx, just a few blocks away from Yankee Stadium, Abu Talib and a group of local residents have been growing food for nearly two decades on what was once a trash-strewn lot. A short subway ride away, on a pocket of leftover land between the Metro North rail line and the Major Deegan Expressway, at one of the borough's newest urban farms—La Finca del Sur (Spanish for "Farm of the South")—volunteers and supporters are monitoring the effectiveness of sunflowers and other plants that remove toxins from the soil. Last year, La Finca was the first stop on an inaugural Urban Farm Tour that took New Yorkers in a refurbished trolley car to Bronx-based urban agriculture sites.

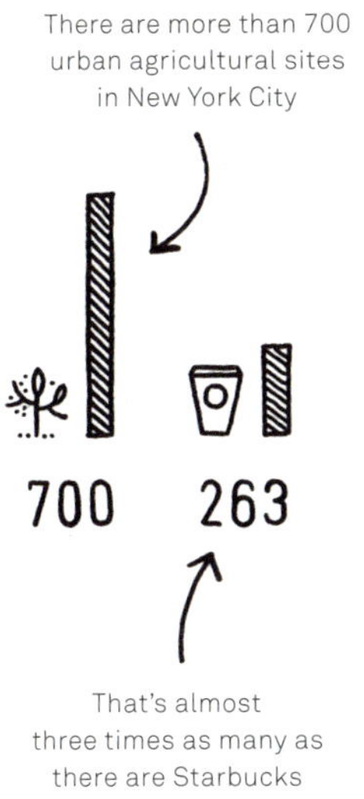

On Manhattan's far west side, students help raise 7,000 pounds of tilapia a year as part of the aquaponics operation at the Food and Finance High School, which is also where City Council Speaker Christine Quinn chose to release FoodWorks, her 2010 policy report which recommends, among other things, increasing the city's food production. Across town at 29th Street and FDR Drive, on one of the city's more than 600 stalled development sites, a temporary farm is providing a nearby restaurant with vegetables grown in 3,600 milk crates. In an industrial district that spans parts of the Greenpoint neighborhood in Brooklyn and Long Island City in Queens, the city's newest commercial rooftop farm, Gotham Greens, supplied part of its first harvest last year to Whole Foods, while a relative veteran of the business (Brooklyn Grange, two years old) won a grant from the NYC Department of Environmental Protection to start a new rooftop farm in the Brooklyn Navy Yard that will help manage more than a million gallons of stormwater a year. On the northern shore of Staten Island, the Snug Harbor Cultural Center and Botanical Garden planted crops last year for its new Heritage Farm, bringing agricultural use back to land that was last farmed a half century ago.

WHAT IS URBAN AGRICULTURE?

Urban agriculture can be defined as growing fruits, herbs, and vegetables and raising animals in cities, a process that is accompanied by many other complementary activities such as processing and distributing food, collecting and reusing food waste and rainwater, and educating, organizing, and employing local residents. Urban agriculture is integrated in individual communities and neighborhoods, as well as in the ways that cities function and are managed, including municipal policies, plans, and budgets.

While this definition encompasses an extremely wide range of growing spaces and practices, *Five Borough Farm* focuses on public, communal, and institutional projects. These projects often function as public space, and typically engage a broader range of stakeholders—including City officials, funders, support organizations, and the general public—than individuals growing vegetables on their own rooftops or fire escapes.

This publication uses the term "farmers and gardeners" as shorthand for people growing food in public, communal, and institutional spaces, and participating in the many other aspects of urban agriculture. Hundreds of gardens citywide do not produce food, but provide open space, gathering places, and sites for growing flowers and trees. *Five Borough Farm* focuses on sites that produce food because food production makes a unique contribution to the city, and involves a specific set of policies and programs. However, the project recognizes that all of these spaces contribute in important ways to New Yorkers' quality of life.

City "farmers" cultivate thriving garden, Manhattan, 1973
Photo: Suzanne Szasz

These days, it seems, all eight million New Yorkers' stories could be about urban agriculture. **There are now more than 700 urban agricultural sites citywide.** The number and diversity of new projects emerging across the city, and the groundswell of popular interest, almost obscure the fact that New York City is not necessarily a place you would expect urban agriculture to thrive. It is one of the most densely populated cities in the country, with some of the highest real estate values. It lacks San Francisco's year-round mild climate, or the vast swaths of vacant land found in Baltimore, Cleveland, or Detroit. Yet several factors enable New York City to be a leader in the practice of urban agriculture.

First, today's urban agriculture movement is literally rooted in the past. Food production has always been part of New York City's landscape, particularly during times of war or economic turmoil in the past century. In the 1970s, as hundreds of thousands of New Yorkers moved out and housing abandonment soared, residents began transforming some of the city's more than 10,000 vacant lots into community gardens. The infrastructure created in that era—including city-sponsored programs and land transfer mechanisms, a strong network of nonprofit advocates and support organizations, and a passionate, resourceful community of activists and garden enthusiasts—has helped make the current urban agriculture renaissance possible.

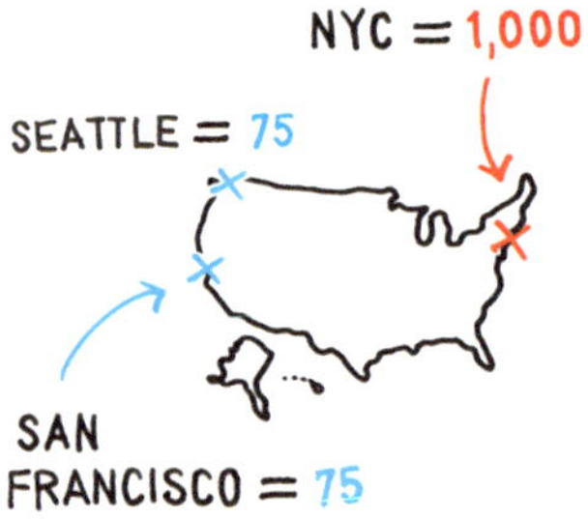

Compared to other municipalities nationwide, New York City has some very supportive urban agriculture–related policies and programs. While many cities have spent the last few years wrangling over zoning code changes to recognize urban agriculture as a permissible land use, New York City's zoning ordinance has allowed gardening and farming virtually everywhere.[6] **The Department of Parks & Recreation and the New York City Housing Authority run two of the largest community gardening programs in the country, with more than 1,000 gardens throughout the five boroughs, most located on public land.** (By comparison, Seattle and San Francisco each have about 75 publicly-owned gardens.)[7] The NYC Department of Education and the nonprofit GrowNYC support an estimated 300 school gardens, 117 of which grow food, and a farm-to-cafeteria program in over 50 schools that incorporates student-grown produce in the school's lunches.[8] The City's elected officials have taken note of the momentum surrounding urban agriculture: in addition to the City Council Speaker's FoodWorks plan, the Mayor's recent update of

6 New York City zoning allows "truck gardens" in residential and commercial zones provided that no offensive odors or dust are created and that only products produced on the same property are sold from the garden. Urban agriculture is permitted in New York City's manufacturing districts (M1, M2, or M3) without these nuisance or sales restrictions. See City of New York Zoning Resolution (web version) Article II: Residence District Regulations. Chapter 2 – Use Regulations. Section 22-14. Effective Date: 4/22/09

7 See http://www.seattle.gov/neighborhoods/ppatch/; http://sfrecpark.org/CommunityGardens.aspx; and SPUR. Public Harvest: Expanding the Use of Public Land for Urban Agriculture in San Francisco. February 15, 2012, pg. 10

8 See http://growtolearn.org/view/RegisteredSchoolGardens for a list of registered school gardens. The Garden to School Café project is a collaboration among the NYC Department of Education's SchoolFood program, the NYS Department of Agriculture & Markets, and the NYC Department of Parks & Recreation GreenThumb program.

La Finca del Sur, Bronx

Billy Ortiz, Berry Street Garden, Brooklyn

the City's sustainability strategy, PlaNYC, and the Manhattan Borough President's white paper FoodNYC endorse the growth of urban agriculture.

New York City also boasts an extensive network of technical assistance providers, advocacy organizations, philanthropies, and private businesses that supports farmers markets (enabling New Yorkers to buy food grown in the city's gardens and farms), organizes neighborhood-scale composting projects, designs and builds gardens and chicken coops, lends gardening tools, trains farmers, and lobbies for supportive public policy. Within the past few years, the city's urban agriculture community has successfully gained the right to raise chickens and bees,[9] and secured new review procedures to strengthen protections for community gardens. Groups have launched an urban farm school to train a new cohort of farmers, and have begun planning a farm incubator program at Floyd Bennett Field, an abandoned airfield a few minutes' drive from Coney Island.

Common to all of these ventures are some very familiar New York City themes: ambitious, resourceful individuals are developing innovative ventures, buoyed by a critical mass of like-minded supporters and a network of services that are most likely to thrive in the hothouse of big, dense cities. As one commercial farmer explained, **"New York City is not the most likely candidate for urban agriculture at all in my mind... what's really interesting and unique is that New York City is a city full of entrepreneurs and people who make smart connections."**

Fresh, homegrown vegetables, herbs, fruits, and eggs can be extremely important to individuals, families, and communities. But urban agriculture is also about much more than food. The city's gardens and farms beautify neighborhoods, provide much needed open space, and serve as safe places for seniors and children to socialize. As permeable green spaces, they reduce stormwater runoff and summer temperatures. Students learn about ecology and food systems at school gardens, while teenagers earn income as they gain skills managing farmers markets and other programs. And urban agriculture is often a

9 Department of Health and Mental Hygiene. 24 RCNY Health Code § 161.01 (2010); NYC Health Code Article 161 requires persons keeping non-aggressive honeybees (Apis mellifera) to file a notice with the Department of Health and Mental Hygiene containing the beekeeper's contact information and location of the hive.

Urban agriculture is clustered in areas of the city that have relatively high concentration of vacant lots as well as pressing social, health, economic, and environmental issues.

Community Gardens & Vacant Lots

■ Community garden locations

\# vacant lots

- Low
-
-
-
- High

SOURCE: GrowNYC 2011 and MapPLUTO
NYC Department of City Planning

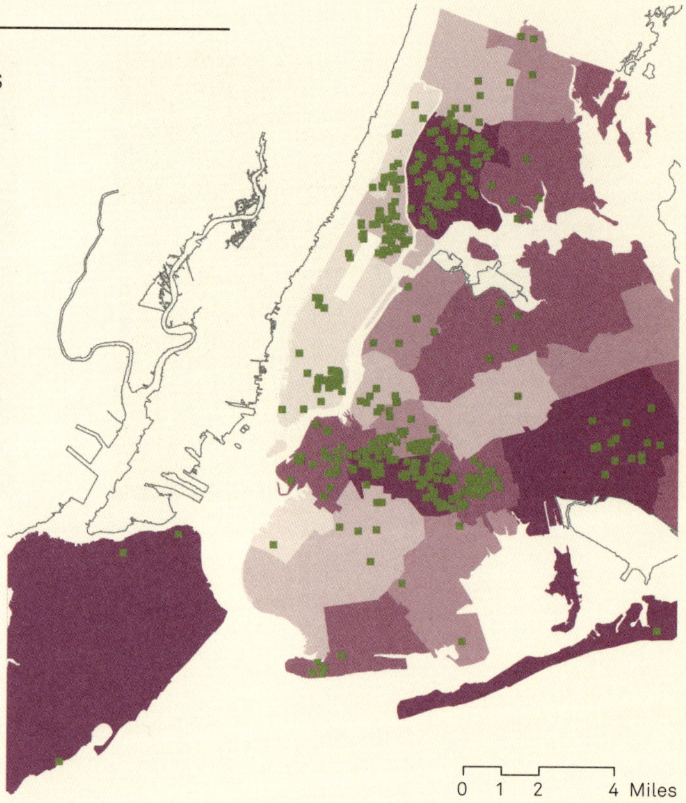

Community Gardens & Vegetable Consumption

■ Community garden locations

% survey respondents ate no fruits or vegetables the previous day

- 1%–5%
- 5%–10%
- 10%–15%
- 15%–20%
- 25%–30%

SOURCE: GROWNYC 2011 and Bureau of Epidemiology Services, NYC Dept. of Health and Mental Hygiene Community Health Survey 2000. Boundaries on the map define 42 United Hospital Fund (UHF) neighborhoods, which are clusters of zip code areas.

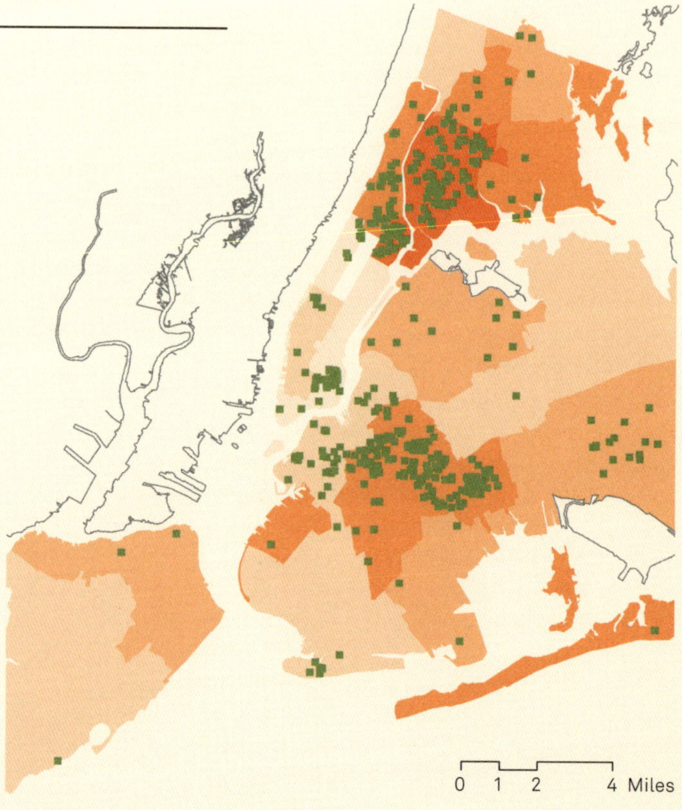

vehicle for community organizing and social justice campaigns, particularly in the city's poorest communities.

It is in these neighborhoods that urban agriculture is arguably making the greatest impact. For although interest in urban agriculture cuts across all demographic and geographic lines, the city's farms and gardens are clustered in places that were hardest hit by decades of disinvestment: Bedford Stuyvesant and Brownsville, the South Bronx and East Harlem. Residents in these areas face a familiar litany of challenges: limited access to healthy food options, underperforming schools, high incidence of diabetes, high unemployment rates, and twice as many vacant lots on average than in the city's wealthy neighborhoods.[10] Urban agriculture gives people one way to turn abandoned parcels of land into a community asset that addresses many pressing needs. As one funder explained: "I really [have seen] a reframing of proposals, from that deficit-based 'we don't have this, we have all these terrible things' to 'we have this potential strength in all these things that can be tied into urban agriculture.'"

Vernon Cases Garden, Brooklyn

The many facets of urban agriculture engage the school system and the public health community, neighborhood block associations and City Hall, contributing to a wide range of health, social, environmental, and economic benefits that align with the goals of government agencies. In recent years, government officials nationwide have embraced the concept of "green infrastructure"—open spaces that serve multiple functions such as providing recreation opportunities, capturing stormwater, and beautifying neighborhoods. In this context, urban agriculture could be considered a kind of "superinfrastructure"—few other open spaces in the city have the same potential to deliver such a broad range of benefits as an active, well-managed, fully supported urban farm or garden.

Yet while these benefits have been well-documented in scholarly literature,[11] they have not been measured systematically in New York City, making it difficult for policy makers, philanthropic organizations, and advocates to substantiate and quantify the impacts of urban agriculture citywide. Partly as a result, the many government agencies that currently deal with the city's farmers and gardeners have limited authority and resources to work with other agencies or incorporate urban agriculture into potentially complementary plans and operations. To address these gaps, *Five Borough Farm* uses photographs, maps, infographics, and results from interviews with dozens of leading urban agriculture stakeholders citywide to provide a detailed portrait of existing urban agricultural activity in New York City. In addition, the project establishes a methodology to track and evaluate urban agriculture's myriad impacts, and provides a road map for City officials, working with private and nonprofit partners, to identify specific initiatives that can support urban agriculture citywide, and amplify the tremendous grassroots efforts taking place every day in all five boroughs.

10 For analysis of the prevalence of vacant lots citywide, see Columbia Urban Design Lab. "The Potential for Urban Agriculture in New York City," p. 55.

11 See, for example, Draper, C., & Freedman, D. (2010). Review and Analysis of the Benefits, Purposes, and Motivations Associated with Community Gardening in the United States. Journal of Community Practice, 18(4), 458-492.

Project Origins

In 2009, the Design Trust for Public Space issued an open call for projects to improve public space in New York City, receiving twenty proposals from organizations citywide. The nonprofit organization Added Value and the design firm thread collective submitted a proposal to address urban agriculture. This proposal was one of two selected by an independent jury of architects, policy experts, open space advocates, and Design Trust board members for the Design Trust to take on as a project.

Project Goals

Five Borough Farm is a project of the Design Trust for Public Space, in partnership wit Added Value, which operates the 3-acre Red Hook Community Farm in Brooklyn. The project has three main goals:

1. Survey and document New York City's existing urban agricultural activity using photographs, maps, and infographics.

2. Establish a shared framework and tools to allow users to track urban agricultural activities citywide, and evaluate their social, health, economic, and ecological benefits.

3. Develop policy recommendations that will help make urban agriculture a more permanent part of the city's landscape and governance.

Project Team

The Design Trust commissioned an interdisciplinary team of private-sector Project Fellows to lead the project work, including experts in food policy, sustainable design, and public health, a graphic designer, and a photographer. The Project Fellows and Design Trust staff engaged key stakeholders to shape and inform the Project, through a large-scale participatory workshop, small advisory groups, and scores of one-on-one interviews; used several different media to communicate findings to a broad, diverse audience, from policy experts to farmers and gardeners; and produced recommendations and tools that could be used to effect concrete change in the City's urban agriculture policies and practices.

Methods

The Design Trust team conducted extensive research and outreach to gather and synthesize information about urban agriculture in New York City. This outreach and research process included the following:

Advisory Committee[12]

To develop the project goals and scope, identify candidates for Project Fellowships, and provide ongoing feedback on developing recommendations and tools and structuring outreach, the Design Trust assembled an advisory committee composed of key urban agriculture stakeholders.

Citywide Workshop[13]

In December 2010, the Design Trust convened a citywide urban agriculture workshop: "Five Borough Farm: The Future of Farming in NYC." More than 90 farmers and gardeners, support organization staff, and funders gathered to share information about their goals and priorities, needs and challenges, and current and desired methods for tracking the impacts of their work. The Design Trust synthesized findings from this workshop in a report distributed to all participants in February 2011.

Five Borough Farm Workshop
Photo: Chris Kannen

12-13 Full list of members and workshop participants on p.154

Project Interviews

Following up on the responses from the workshop, the team identified four key urban agriculture stakeholder groups for one-on-one, in-depth, confidential interviews. These groups included:

Farmers and gardeners: Urban gardeners and farmers who produce food for personal or community consumption, including for market customers. These farmers and gardeners represented a diverse mix of operations in terms of organizational structure (see page 160 for working definitions used in this project), location (with representation from all five boroughs), size of production site, leadership demographics (e.g., race/ethnicity; gender), and community served by the farm or garden

Support organizations: Representatives from nongovernmental organizations that provide technical, material, or educational support, and/or advocated for urban agriculture in New York City

Funders: Representatives from private foundations that had recently funded urban agriculture and/or urban food systems programs

Government officials: Representatives from New York City, state or federal agencies directly involved with supporting and/or regulating one or more aspects of urban agriculture activities throughout the five boroughs

From January to June 2011, the Policy and Metrics Fellows conducted confidential interviews with 31 individuals: 16 farmers and gardeners, six government officials, five support organization staff, and four funders. The interviews consisted of a series of open-ended and follow-up questions covering current practices, indicators of success, constraints, policy issues, and recommendations for expanding urban agriculture in New York City. The interviews were recorded and transcribed; the transcripts were then coded and analyzed for consistent themes and differences across the interviews, with quotes organized to illustrate or explain the concepts being discussed. The individual interviewees remain anonymous.

The team also met or spoke with an additional 40 farmers and gardeners, support organization staff, policy makers and funders to address specific questions that arose from the interviews. Findings from the formal and informal interviews and meetings are summarized in the chapter titled "Urban Agriculture in NYC."

In addition to reviewing information from the formal interviews and from other urban agriculture stakeholders, the Policy Fellow engaged in the following activities to develop recommendations:

- Conducted semi-structured in-person interviews with urban agriculture advocates, farmers and gardeners, and policy makers in Chicago, Minneapolis, San Francisco, Seattle, Toronto, and Vancouver, and reviewed policy documents and reports from each city

- Reviewed policy documents from New York City, including published food plans, relevant sections of the City Charter, Administrative Code, zoning text, and planning documents, and reports and documents of urban agriculture support organizations

- Participated in Manhattan Borough President Scott Stringer's food policy task force and City Council Speaker Christine Quinn's FoodWorks advisory group

- Convened a working group of farmers, gardeners, government officials, and policy makers to get feedback on initial recommendations

The Metrics Fellow engaged in the following activities to develop a framework for tracking and evaluating urban agriculture's benefits:

- Reviewed literature to find previous studies that evaluated the impacts of activities common to urban agriculture

- Visited farms and gardens citywide, and reviewed documents and records used to track activities and outcomes, to understand farmers' and gardeners' capacity to track and evaluate their programs

- Convened a working group of farmers and gardeners to get feedback on an initial set of recommended methods to track and evaluate benefits

Design Fellow thread collective contributed research and mapping, and conducted outreach and informal interviews. With project partner Added Value, thread collective also reviewed and helped shape the final publication and website.

The Photo Urbanism Fellow visited and documented more than 100 farms and gardens citywide from June 2011 to June 2012.

Finally, the Design Trust convened two small working groups of urban agriculture stakeholders to review and provide feedback on specific metrics and policy recommendations.

Eagle Street Rooftop Farm, Brooklyn

I. URBAN AGRICULTURE IN NYC

63rd and 64th Street Community Garden, Brooklyn

FedEx

Top: Backyard garden, Brooklyn   Bottom: Temple of David Community Garden, Brooklyn

Garden of Happiness, Bronx

Kissena Corridor Community Garden, Queens

Top: East New York Farms, Brooklyn   Bottom: Randall's Island Learning Garden

Hell's Kitchen Farm Project, Manhattan

Top: Garden, Beach 90th Street, Queens   Bottom: Seeds to Feed Rooftop Farm, Brooklyn

Bruckner Mott Haven Garden, Bronx

Eagle Street Rooftop Farm, Brooklyn

Kissena Corridor Community Garden, Queens

Top: Taqwa Community Farm, Bronx   Bottom: Culinary Kids Garden, Queens

INTRODUCTION

Urban agriculture comes in many shapes and sizes. One of its most fundamental units in New York City is a raised bed—typically a box made of wood filled with soil, which allows people to grow in healthy soil above the ground.[14] Most raised beds measure three by five feet or four by eight feet, a size that is manageable for almost anyone to cultivate, and that can produce surprisingly large yields.

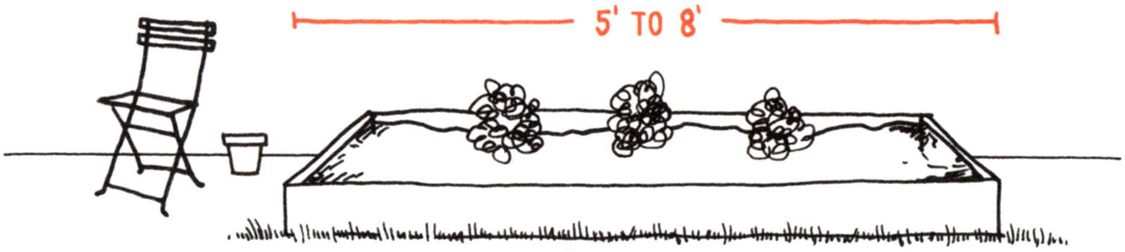

Zoom out to the scale of a single garden or farm. Volunteers and local residents are tending vegetables, fruit trees, and livestock, and also turning food scraps into compost, or leading students from a nearby elementary school in an after-school program. Others are learning to cook a vegetable they have never tried before or attending a community workshop about the federal Farm Bill. Residents from around the neighborhood stop by to pick up their Community Supported Agriculture share, while deliveries of produce go out to local churches, food pantries, restaurants, or bodegas.

14 Because urban soil may contain contaminants from previous uses, many urban farmers and gardeners growing food avoid planting directly in the ground.

Zoom out again to the scale of the city, and urban agriculture appears as a complex system, with thousands of raised beds in hundreds of schoolyards and community gardens, on rooftops and public housing land. Scores of community-based organizations, government agencies, and philanthropies contribute soil, lumber, and funding, and run horticultural training and nutrition classes. Upstate and Long Island farmers and suppliers bring food and seeds, connecting city residents to the surrounding region. Underlying all of this activity are the invisible laws, regulations, and policies that influence where new farms and gardens locate and what activities are encouraged or prohibited.

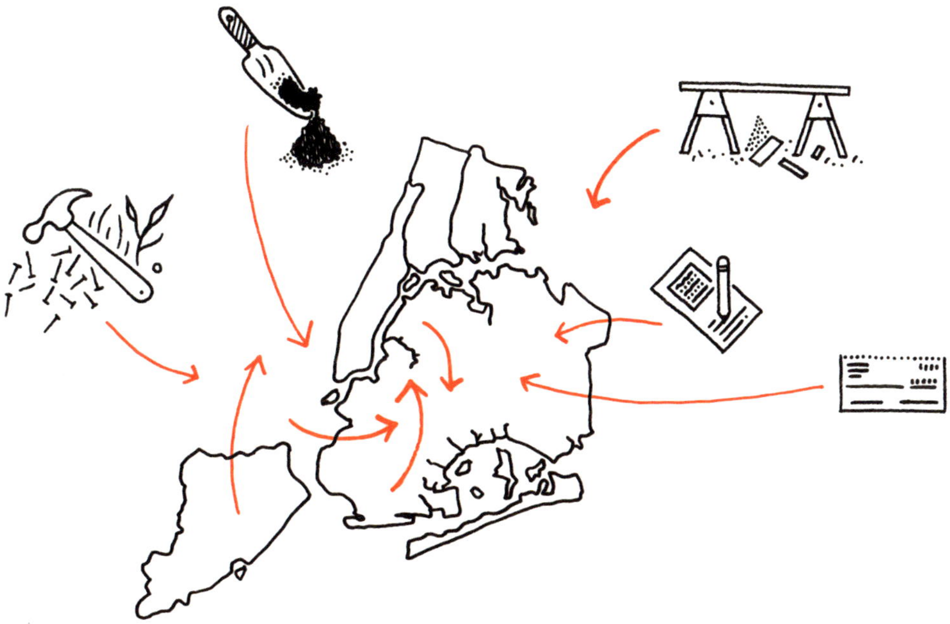

In order to recommend lasting and meaningful improvements to the urban agriculture system, and to produce tools that will be useful to its diverse spectrum of actors, the first step is to understand how it functions. This section provides a brief overview of the city's urban agriculture system, including:

- Goals cited by farmers and gardeners, and the activities and programs they offer to meet those goals

- Defining characteristics of four main types of urban agriculture operations in New York City: institutional farms and gardens, commercial farms, community gardens, and community farms

- Resources the city's farmers and gardeners need to operate, and the challenges they face in obtaining those resources

- Roles of three other key urban agriculture stakeholder groups in New York City: support organizations, government officials, and funders

URBAN AGRICULTURE

Urban agriculture involves many different types of food-producing spaces, stakeholders, resources, and policies, and contributes to many benefits.

Health

Access to healthy food

Food-health literacy

Healthy eating

Physical activity

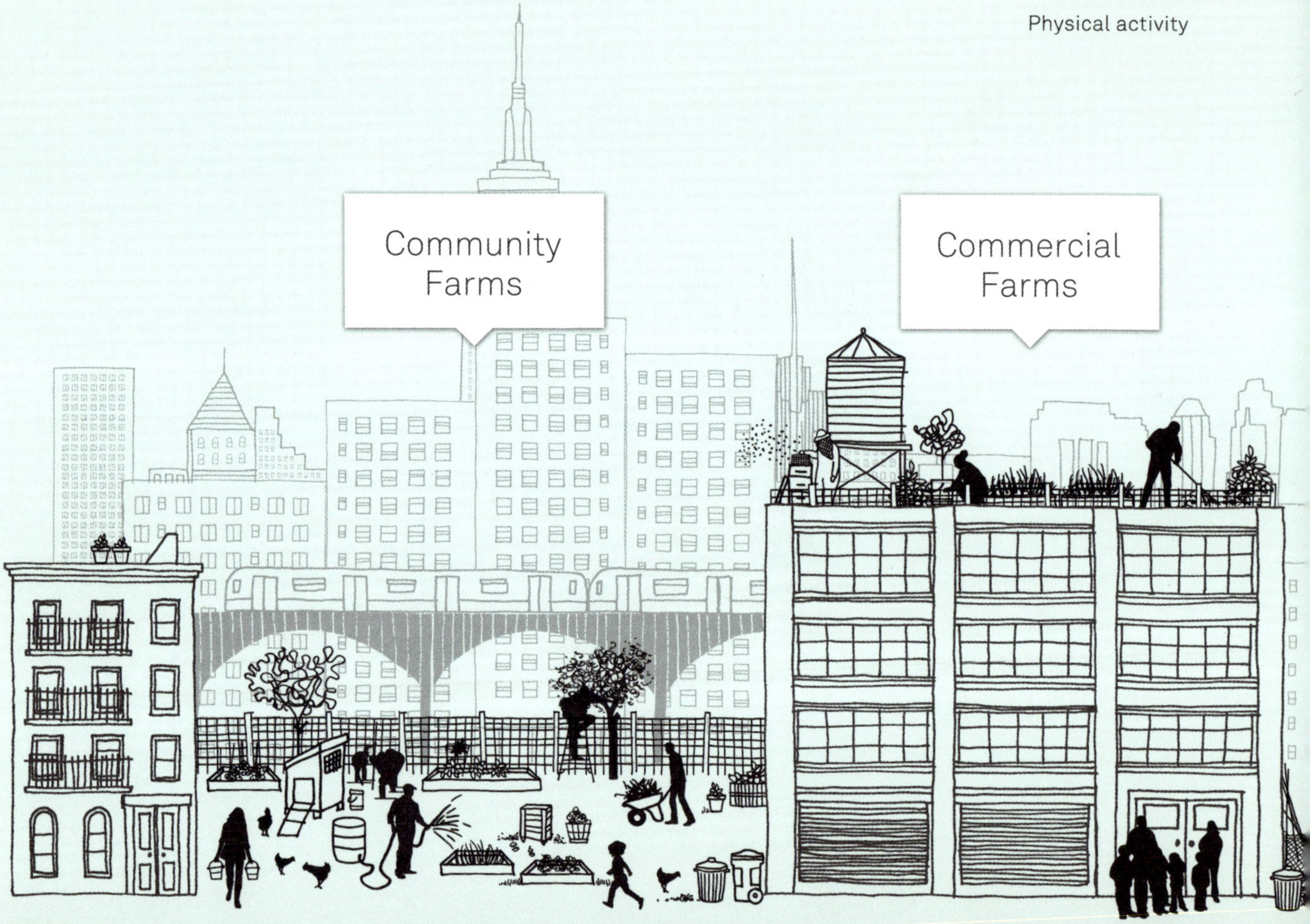

People

Local residents

Volunteers

Community organizations

Students

Visitors

Market customers

Soil & Compost

Self-produced

Purchased

Donated

Supplies

Seeds

Fertilizer

Tools

Construction Materials

Water

Electricity

Benefits

Social

Empowerment + Mobilization

Youth Development & Education

Food Security

Safe spaces

Socially Integrated Aging

Economic

Local economic stimulation

Job Growth

Job Readiness

Food Affordability

Ecological

Awareness of Food Systems Ecology

Stewardship

Conservation

Storm Water Management

Soil Improvement

Biodiversity + Habitat Improvement

Key Stakeholders

Farmers and Gardeners

Government Officials

Support Organizations

Funders

Financial Resources

Sales of produce

Grants

Donations

Fees for services

Support Services

Technical assistance

Advocacy and policy work

Environmental education

Networking events

Access to land and rooftops

GOALS & ACTIVITIES

Some New Yorkers are involved in urban agriculture simply because they want to grow their own fresh vegetables. But farmers and gardeners of all varieties describe a broad and civic-minded set of goals and objectives that guide their work, including:

- Educating young people, and improving their job readiness and leadership skills
- Providing safe spaces for local residents
- Cleaning and remediating vacant or underused spaces
- Connecting residents to nature
- Creating opportunities for residents to earn income

To these farmers and gardeners, growing food is important, but often food production is described as a means to other goals as well. Many of these goals align with the missions of government agencies and philanthropic organizations, even if these entities are not explicitly collaborating with farmers and gardeners on their projects.

This section, while not comprehensive, highlights some of the activities that take place at the city's farms and gardens, and organizes them under four broad categories:

- **Health**
- **Social**
- **Economic**
- **Ecological**

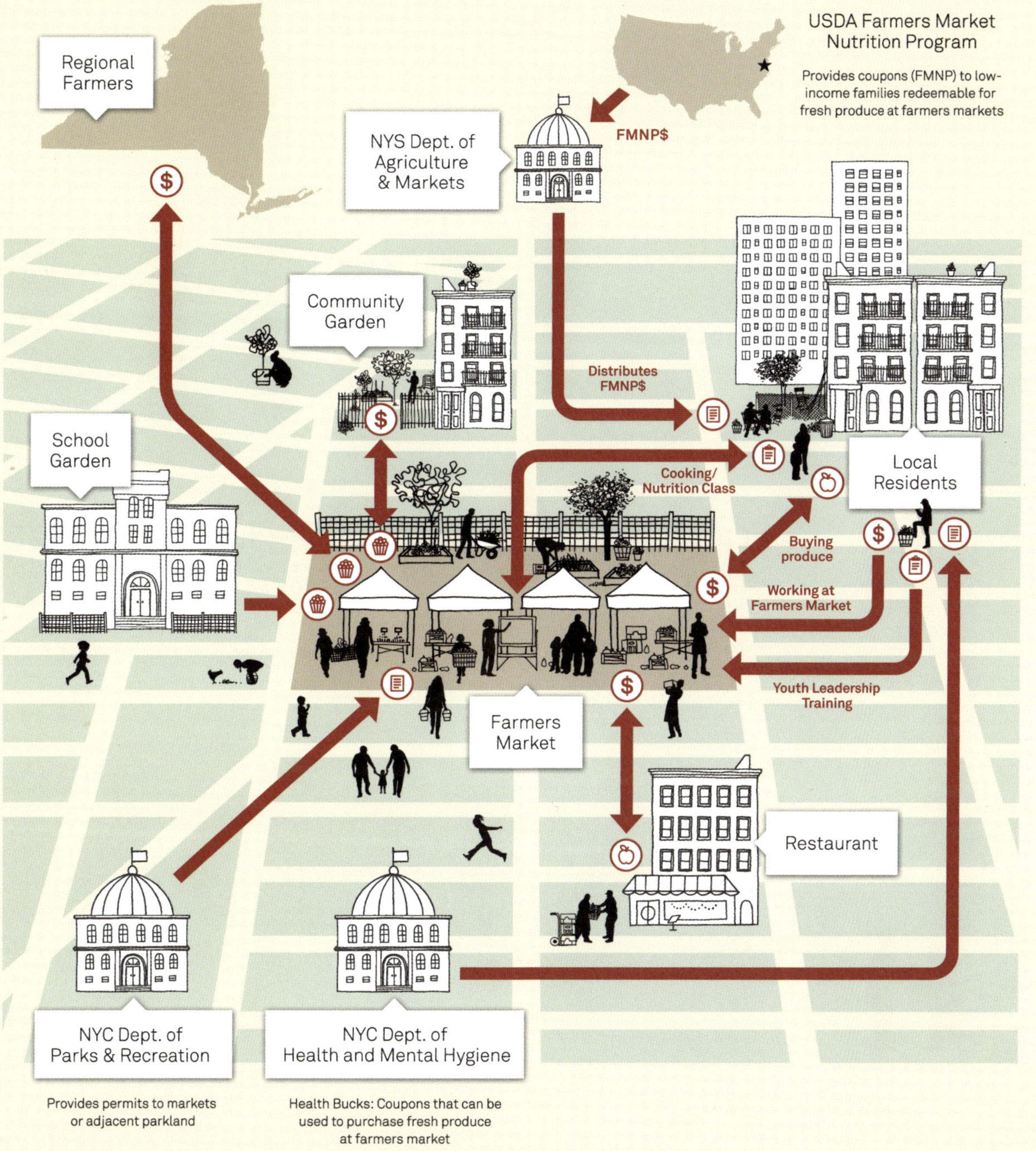

FARMERS MARKET

A single urban agricultural activity can contribute multiple benefits, and involve many individuals and institutions. For example, a number of the city's farms and gardens participate in a farmers market. Growers earn income selling produce. Teenagers develop job skills as paid interns managing the farm. Local residents benefit from having a nearby option for fresh vegetables and from programs that offer cooking and nutrition classes. These outcomes overlap with the goals of a number of municipal agencies, funders, and support organizations.

- Supplies produce
- Provides access to produce
- Earns income
- Provides subsidy or permit
- Provides education or training

Benefits:

- Access to healthy food
- Food-health literacy
- Healthy eating
- Empowerment & mobilization
- Local economic stimulation
- Job growth
- Job readiness
- Food affordability
- Conservation

Health

Urban farmers and gardeners seek to improve the health of participants through a variety of strategies: providing access to nutritious food, which is particularly important in communities with few retail establishments selling fresh produce; educating people about the relationship between good nutrition and health; motivating people to increase their consumption of fruits and vegetables; and involving people in gardening as a form of physical activity.

"If people grow their own food, or see where their food comes from, or buy food at farmers markets where fruits and vegetables taste so much better, they get much better value, and they may be more likely to eat healthier and continue eating healthier when the farmers market is out of season."

– Government Official

87% of the city's community gardeners responding to a 2009 survey reported eating vegetables they grew, and about half reported donating produce.

About 10% of the city's gardeners reported they sell produce at farmers markets.

Distributing food

The food grown in gardens and farms reaches people in numerous ways. Many people prepare the food they grow for their families. Growers also donate food to farm or garden visitors, or to food pantries, churches, and other institutions. Some farms and gardens sell the food they produce to customers through farmers markets and Community-Supported Agriculture (CSA) programs. In some cases, farms and gardens sell produce at nearby bodegas, supermarkets, or to local restaurants.

Community-Supported Agriculture (CSA)

CSA programs allow consumers to buy produce at the beginning of the growing season, receiving weekly allotments directly from farms. Nearly 100 Red Hook residents participate in Added Value's CSA, the first in the city to provide members with weekly shares of produce grown in their own neighborhood. All members provide at least nine hours of sweat equity on the farm each year; some shares are reserved for low-income residents.

Cooking and nutrition classes

Farmers, gardeners, and organizations and institutions, including nonprofits and City agencies such as the Department of Health, provide classes to teach people about the benefits of healthy diets and how to prepare foods found at farmers markets. For instance, one community farm trains local residents to give cooking demonstrations and presentations on nutrition at senior centers, churches, and other neighborhood institutions.

"We give people the excitement, the motivation, the impetuousness, the guilt, whatever it takes [to make healthy eating decisions]."

– Commercial Farmer

Social

Urban farms and gardens offer spaces for people to meet, beautify the neighborhood, and join campaigns for social or political change. One community gardener referred to the garden as a "community resource center."

Clearing and transforming vacant land

One of the first steps in creating an urban farm or garden is often clearing an empty lot. Local residents can get assistance from City agencies in removing and trucking away garbage and debris.

"[If] you have a site that's abandoned and people are dumping garbage there and then it's taken over for gardening, it's not going to look like a garbage dump.... And so by extension, if it's used for farming... then it's

going to be by definition safer. We won't have antisocial activities going on there, [and] people won't become victims of crimes..."

– Government Official

Social gathering spaces

Farms and gardens host a broad range of public events, from movie screenings to barbecues to harvest festivals, that allow people to gather—whether or not they are involved with growing food.

Intergenerational interaction

Urban agriculture sites typically attract residents and volunteers from a broad range of age groups. Some sites offer programming specifically designed to foster interactions among different generations: one community gardener hires teenagers to distribute food from the garden to local seniors who are unable to leave their homes.

Red Hook Community Farm, Brooklyn

"[Urban agriculture is about] meeting your neighbor, having seniors interact with young people, [enabling people to] understand that we're not isolated and living in silos."

– Funder

Community-based research and organizing

One community gardener described a partnership between local residents and Columbia University urban planning students to survey vacant lots in the garden's vicinity, and develop criteria to assess the suitability of these lots for urban agriculture.

Food justice/social justice education

Numerous farmers and gardeners said one of their goals is to educate participants about inequalities in the food system and in wealth distribution, and to connect that information to their neighborhoods. One community farmer, describing the garden's annual youth education and leadership program, said:

> There are some things that are just imperative that frankly they don't get in the schools they go to [such as] race and culture and wealth, like concepts of wealth, what does it mean to be wealthy? And connecting health to food access. Different kinds of things that are important to this movement that we have to talk to them about explicitly or else they're not going to make the connection on their own in the same sort of way.

"Urban agriculture has really been a good way to get people into leadership, and once they're in leadership, to get them to understand the various forces around land use and zoning and development and those kinds of issues, and then be able to mobilize around an advocacy agenda... That is important for us because we believe in order for [positive gains] to be sustained... you need to have a broad base of informed and engaged people supporting that agenda for the long haul."

– Funder

Women-focused programs

Several community gardeners have explicit goals to empower women in their neighborhoods. "A part of our mission is that we're run by black and Latina women," said one gardener. "[We] provide opportunities for leadership and expression and safety and health for those groups."

Economic

Some farms and gardens host farmers markets, which sell produce from the local urban agriculture site and from regional farmers. Local residents may earn income selling produce they've grown at these markets, or by helping to manage some aspect of the market. Some farms and gardens pay stipends to teenagers to manage farm operations and participate in leadership programs.

Selling food

Local residents who grow food in their backyards or in local community gardens can sell this produce at the farmers markets at Brooklyn's East New York Farms; these growers earn up to $4,000 annually from the markets.

High School for Public Service Youth Farm, Brooklyn

Youth empowerment and job training

Numerous farms and gardens hire teens to manage farms or run markets. In these positions, young people learn general job-readiness skills, as well as skills that can be applied to agricultural, environmental, and food careers.

Employment

Commercial farms, community farms, and institutional farms all typically hire staff. Community gardens selling food at farmers markets sometimes hire local residents to help set up or run the markets.

Ecological

Farmers and gardeners work to improve the environmental conditions of their communities in numerous ways. For instance, since soil amendments and water are often scarce and expensive, urban farmers and gardeners capture and reuse rainwater, and turn food waste into compost. While these practices help people grow food, they also reduce the amount of rainwater that floods the city's sewer systems during storms and the amount of trash the city pays to haul to landfills. These activities also engage local residents as more active stewards in maintaining neighborhood open spaces, and build awareness of infrastructural systems (e.g., stormwater, waste) that impact the city's environment.

Rainwater harvesting

GrowNYC has installed 80 collection systems in gardens citywide, but there is no comprehensive tracking of how many other sites citywide capture rainwater for irrigation.

Composting

In 2009 more than 130 community gardens reported that they currently have composting systems to collect either food waste or yard waste.

Soil remediation

La Finca del Sur, a community farm in the South Bronx, has partnered with the architecture and urban planning firm youarethecity to create Field Lab, an ongoing project to grow, monitor, and harvest several varieties of plants known for their ability to remove toxins from the soil (a process known as phytoremediation).

Environmental education

Through partnerships with local schools and informal visits, urban farms and gardens provide young people with the opportunity to (literally) get their hands dirty and learn about a broad range of topics related to growing plants.

TYPES OF URBAN AGRICULTURE

New Yorkers are growing food wherever space is available: in schoolyards and on the grounds of public housing projects, on privately- and City-owned vacant lots, on rooftops and in public parks. No two growing spaces are alike—they may be maintained by volunteers or paid staff, have budgets of a couple hundred or hundreds of thousands of dollars, and be affiliated with one of a dozen different entities that control the land. Each urban agriculture project arises in response to the particular needs and opportunities of a given community, organization, or site.

Within this diverse mix, there are some important characteristics that differentiate one type of urban agriculture project from another. Some are run by nonprofits, for instance, while others operate as for-profit businesses; some emphasize community development rather than food production.

Institutional Farms and Gardens	Commercial Farms
 Edward A. Reynolds West Side High School	 Eagle Street Rooftop Farm
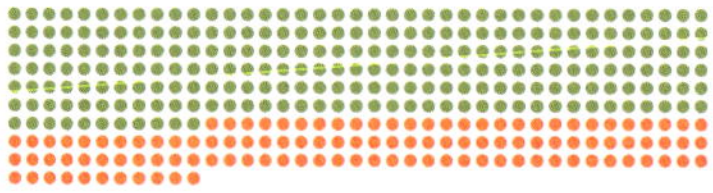 245 NYCHA gardens; 117 public school gardens	3 farms
Affiliated with an institution (e.g. school, housing development, prison, etc.)	Managed as a for-profit business

However, the city's urban farms and gardens also share many similar goals and practices, blurring simple boundaries. A for-profit, commercial farm, for instance, may also run a nonprofit arm that provides free services. Both nonprofits managing farms and community gardeners sell produce at farmers markets, allowing people to earn income from what they grow and generating revenue to support programming. School garden coordinators and public housing residents operate within very different bureaucratic structures, but they all understand the value of providing fresh, healthy food in neighborhoods that often lack these options.

To ensure that a broad range of urban farmers and gardeners were included in the project interviews, the Design Trust initially defined four categories of urban agriculture: institutional farms, commercial farms, community gardens, and community farms. The following section describes the four main categories of growing spaces, highlighting both unique characteristics and common elements.

Community Gardens	Community Farms
 Temple of David Community Garden	 East New York Farms
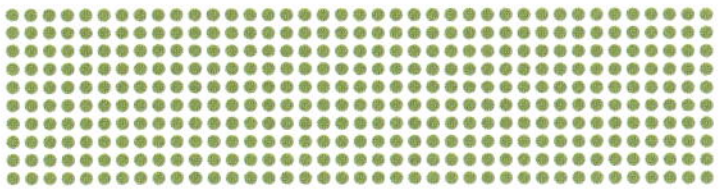 **390** gardens	**7** farms
Typically managed by group of local resident volunteers	Typically managed by a nonprofit organization; includes in its mission engagement with the surrounding community

1. Institutional Farms and Gardens

1a. School Gardens

An estimated 289 New York City schools have active gardens; of these, 117 grow food.[15] School-based urban agriculture comes in many forms. Students grow vegetables on rooftops, in schoolyards, and in indoor facilities, as well as at nearby community gardens and farms. Some schools use formal classroom curricula that incorporate urban agriculture. A handful of schools citywide have extensive urban agricultural programs: for instance, John Bowne High School in Queens has an Agriculture Department with a three-acre farm and an on-site farmers market that enrolls roughly 600 students in an agriculture curriculum.

High School for Public Service Youth Farm, Brooklyn

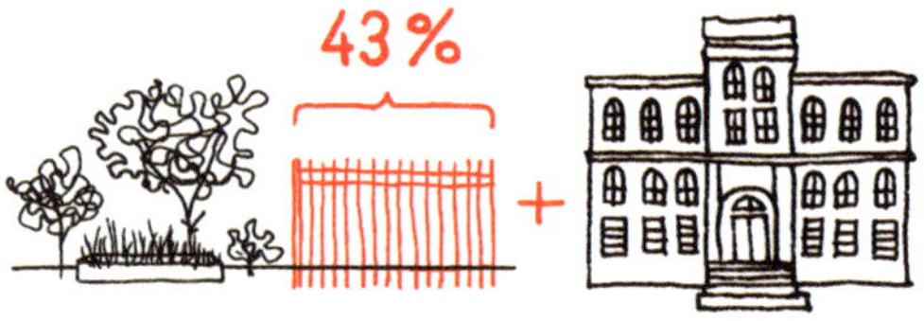

About **43% of the city's community gardens** indicated that they partner with at least one school, hosting educational events and school visits, and maintaining school garden plots.

SOURCE: 2009-2010 GrowNYC Community Garden Survey

15 Based on GrowNYC data from August 2011

School Gardens

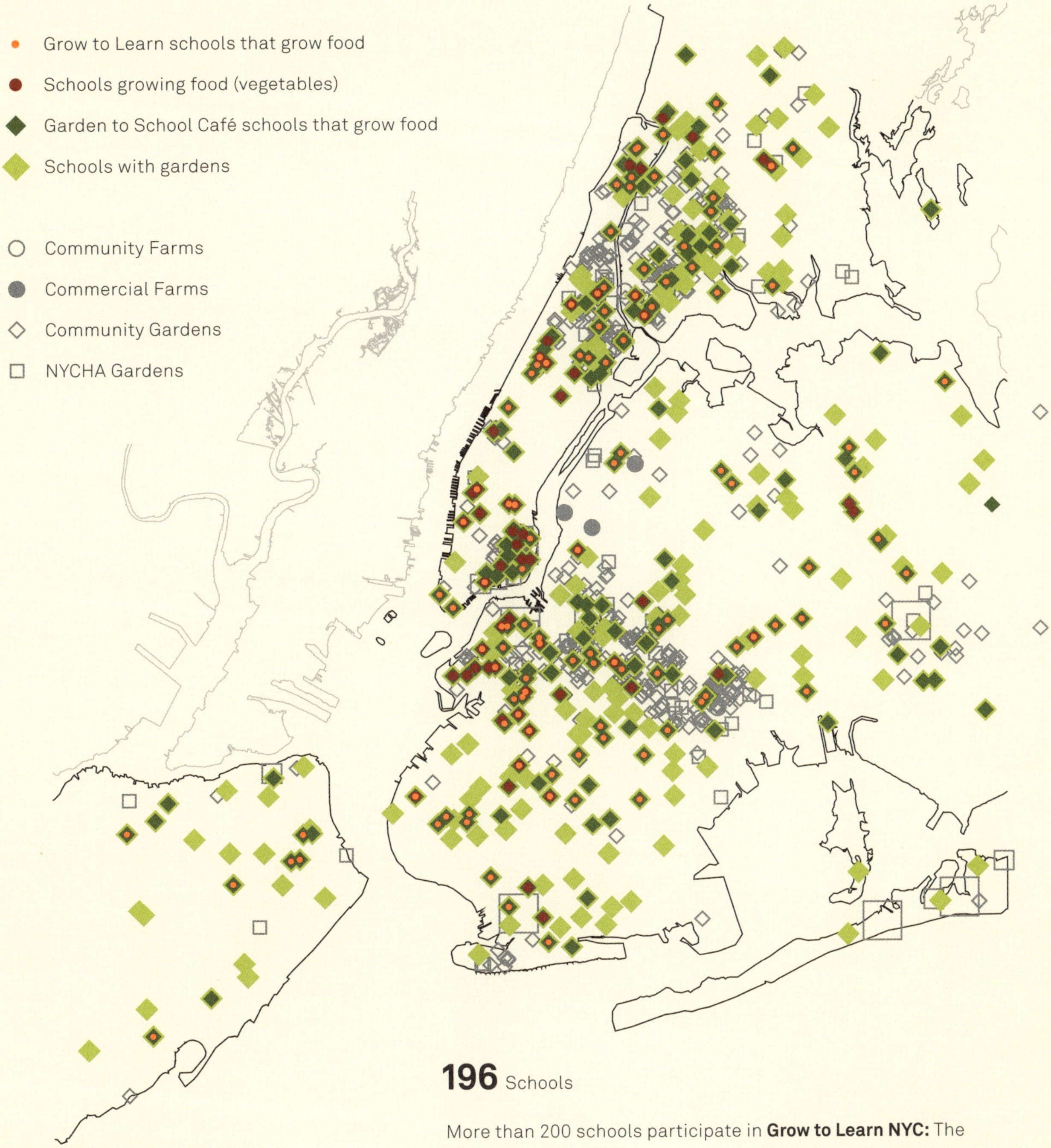

59 Schools

Fifty-nine schools participate in **Garden to School Café,** a Department of Education program that uses student-grown produce in school cafeteria meals.

196 Schools

More than 200 schools participate in **Grow to Learn NYC:** The Citywide School Gardens Initiative, a program established in 2010 by the nonprofit GrowNYC to create and maintain public school gardens. Participating schools are eligible for grants from $500 to $2,000 to start or expand a garden, and are offered free soil and seeds, as well as assistance with lesson plans and curricula for education programs around gardening. To register with Grow to Learn, schools must form a committee of students, teachers, administrators, and parents, define the school's vision for their garden, and create a detailed plan of the garden.

SOURCE: GrowNYC

SARAH FRANK

WEST SIDE HIGH SCHOOL

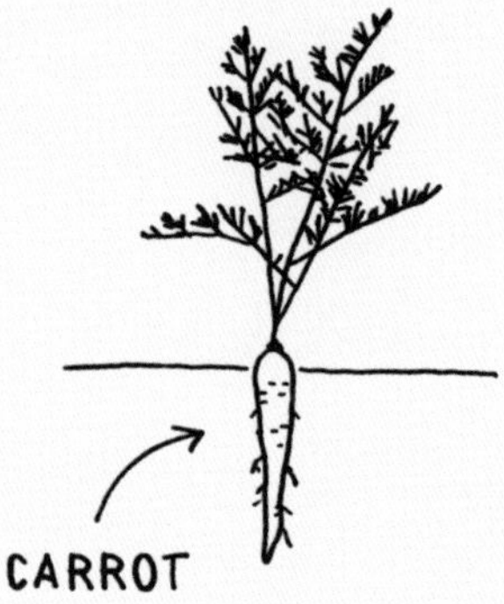

Sarah Frank is a HealthCorpsSchool Coordinator at the Edward A. Reynolds West Side High School at 102nd Street and Amsterdam Avenue in Manhattan. HealthCorps is a national service and peer mentoring initiative.

West Side is an alternative high school - we serve kids who have been struggling in regular high school who need a different type of learning environment. A lot of the hands-on stuff is really the most successful type of activity [for us]. We have a very robust and successful cooking program which is mostly after school, but also some classes cook during the regular school day. We were focusing on vegetables, and we were talking about what it means to eat local food. The garden was sort of a next step.

Many students didn't know that the orange part of the carrot was in the ground. So when I'd bring a class out to the garden, I'd point to the green carrot tops and say, "Well, what is this? Pull one out, let's see what it is." And they would pull it out and be like, "That's mad cute! So the orange part is in the ground?" And I'd say, "Yeah, that's the root."

In terms of yield and education, those two things are often at odds with each other because you can't learn about growing food and be perfect at the same time. So, our feeling is really that this is a place for learning and the productive aspect of the garden is really a secondary goal.

There is a lot of anecdotal evidence about the way that health attitudes have changed in this school. We are really trying to have a holistic approach, so I don't know if I could say it's the garden that caused this. But we have had significant attitude shifts and much increased vegetable consumption in the cooking class. We are seeing this cultural shift. The cooking class I have now is so much further along in being willing to try foods—especially vegetables—in comparison to when I first started here, and I don't think it's chance.

Through an organization called Food Fight, we just launched a semi-school-wide compost program through a private waste management company. Right now we have two composting receptacles in the cafeteria and they get picked up every night. Anything, anything, anything that was ever alive can be composted: paper lunch trays, chicken nuggets, everything. We are pretty excited to be the first public school in NYC to be a part of it.

1b. New York City Housing Authority (NYCHA) Gardens

According to NYCHA staff, the 400,000 residents of the city's 334 public housing developments maintain some 600 gardens; 245 of these grow food.

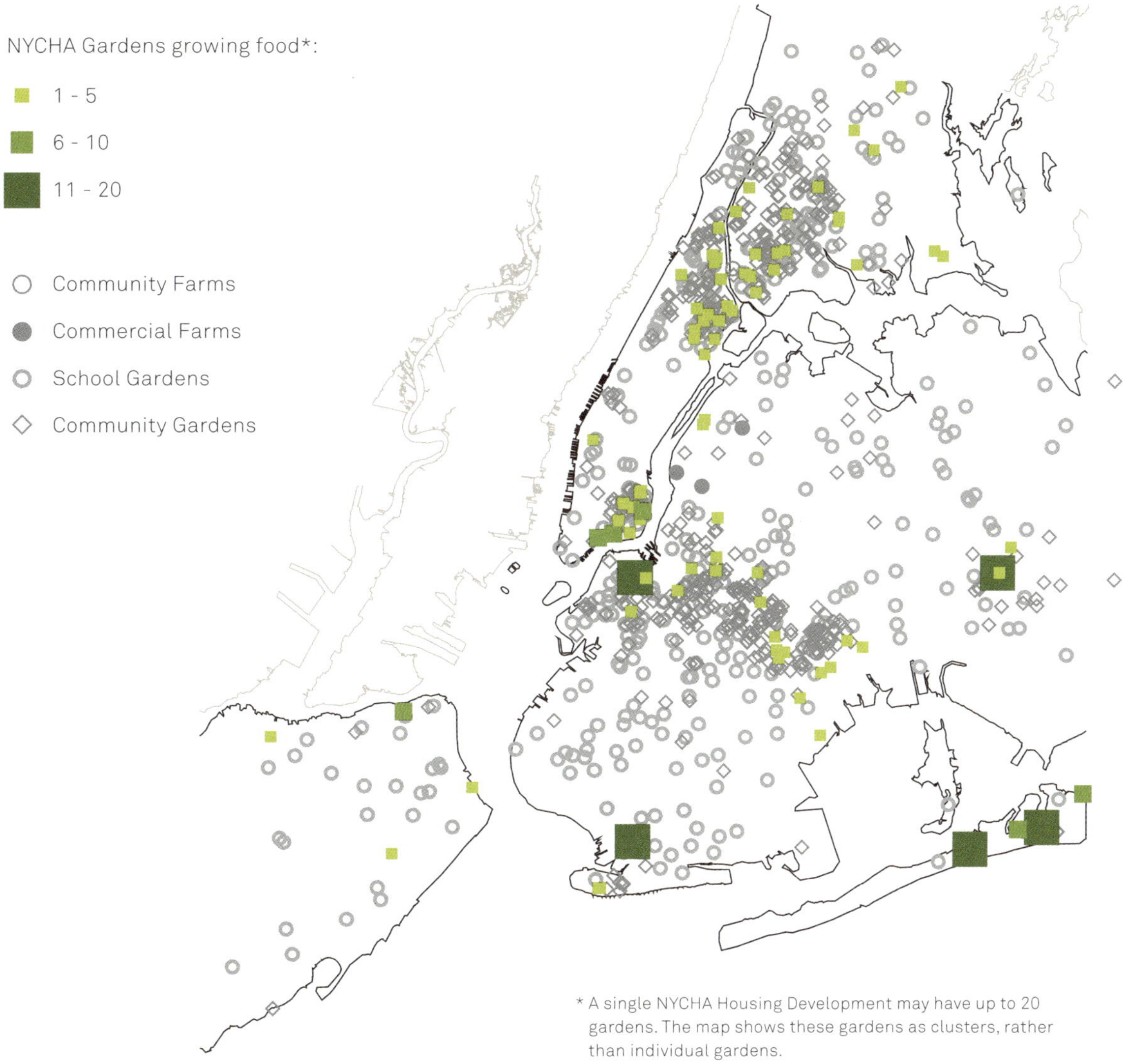

* A single NYCHA Housing Development may have up to 20 gardens. The map shows these gardens as clusters, rather than individual gardens.

SOURCE: NYCHA Garden and Greening Program

The NYCHA Garden and Greening Program provides public housing residents with free seeds and bulbs, tools, technical assistance, labor (such as help turning soil), and reimbursement for $40 of garden-related costs from the NYCHA Property Management Office in each development. In some developments, property management staff source free leaf compost from the Department of Sanitation when it's available.

Seeds to Feed Rooftop Farm, Brooklyn

1c. Other Institutional Farms and Gardens

These include farms and gardens at hospitals, churches, prisons, supportive housing developments, and other institutions whose primary mission is not food production, but which have goals that urban agriculture supports. Examples include:

- **On Rikers Island**, the New York City Department of Corrections, in partnership with the Horticultural Society of New York, maintains a **2.5-acre farm** to provide horticultural therapy and vocational training for inmates.

- **Bowery Mission Rooftop Farm**, on the Lower East Side of Manhattan, produces food that supplements its kitchen, which **serves more than 900 meals daily.**

- **Georgia's Place**, located in the Crown Heights neighborhood of Brooklyn, is a supportive housing facility for the formerly homeless with an **850-square-foot rooftop vegetable and fruit garden.**

2. Commercial Farms

There are a handful of for-profit farms in the city, including three rooftop farms in the industrial areas of Long Island City and Greenpoint: Eagle Street Rooftop Farm, Brooklyn Grange, and Gotham Greens. Later this year, Bright Farms, a private company that develops greenhouses, will begin construction of a 100,000 square foot greenhouse at a City-owned, former Navy warehouse in the Sunset Park neighborhood of Brooklyn. Brooklyn Grange is in the process of constructing a new rooftop farm in the Brooklyn Navy Yard.

In general, commercial farmers try to maximize crop performance in order to achieve profitability, in contrast with community farmers and gardeners. However, some commercial farmers share many of the goals of the broader urban agriculture community, such as promoting healthy eating, contributing to ecological benefits (capturing stormwater, improving air quality, etc.), and providing horticultural therapy. Eagle Street Rooftop Farms, for instance, offers free education programs for grades two and up during the growing season through its sister nonprofit organization Growing Chefs: Food Education from Field to Fork. Brooklyn Grange has partnered with a nonprofit organization, the Refugee Immigrant Fund, to build on the agricultural skills many recent immigrants bring to the city, as well as to provide therapy for victims of torture and persecution. Like community gardens and farms, commercial farms also rely heavily on volunteer labor and donated materials.

Brooklyn Grange, Queens

3. Community Farms

For the purpose of this project, community farms are defined as growing spaces operated by a nonprofit organization (rather than by a group of volunteers, as is typically the case with community gardens) that includes in its mission engagement with the surrounding community. Typically, community farms feature communal growing spaces managed by staff and volunteers, rather than individual plots.

Growing food is an integral part of a community farm's operation. However, unlike commercial farms, community farms focus on community development and social programs such as school education programs and youth leadership training, rather than on maximizing food production and profitability.

Some community gardeners use the term "community farm" to signal that they grow food. For instance, the Prospect Heights Community Farm in Brooklyn is in fact a community garden run by member volunteers.

Added Value, Brooklyn

Community Farms and Commercial Farms

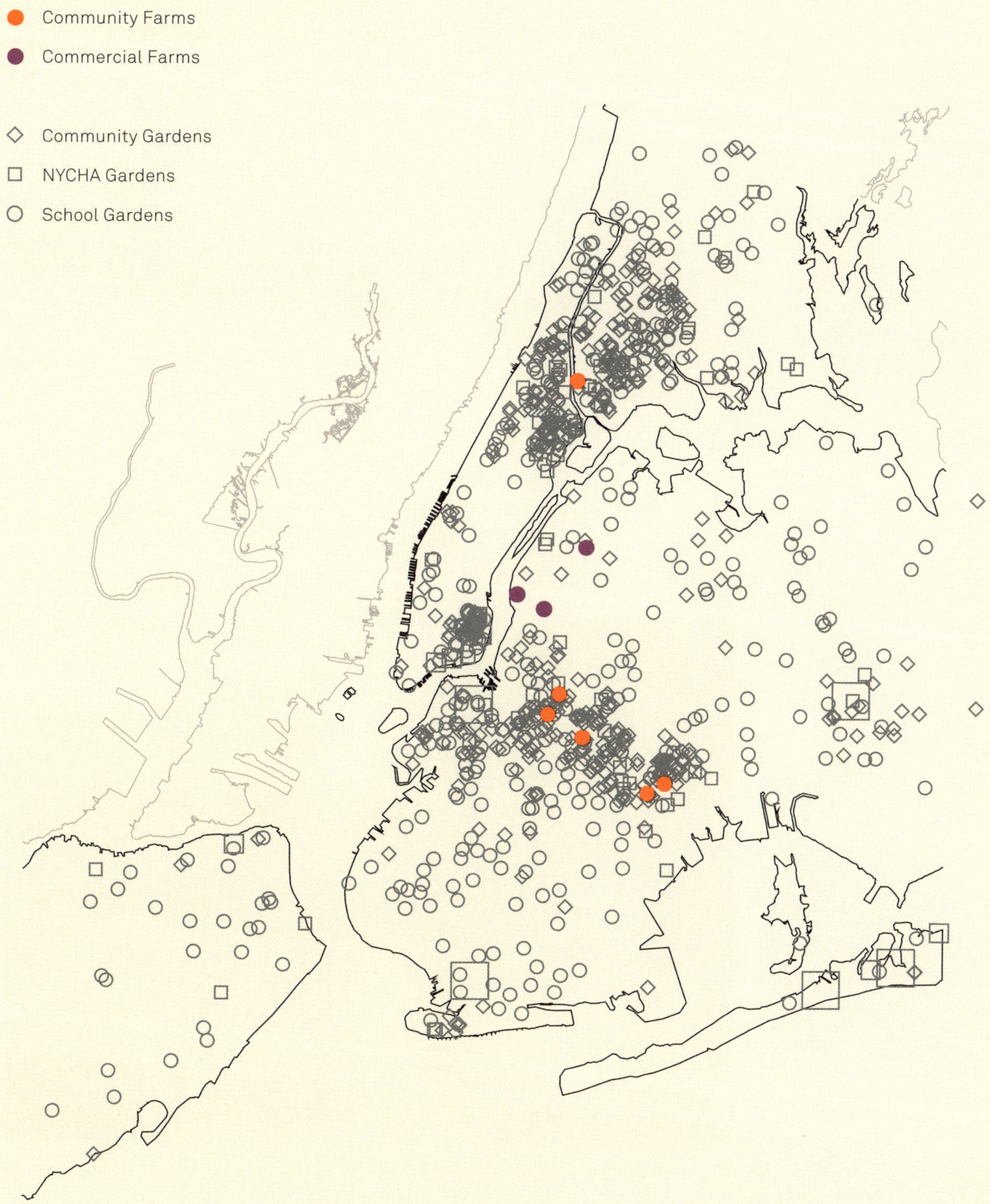

4. Community Gardens

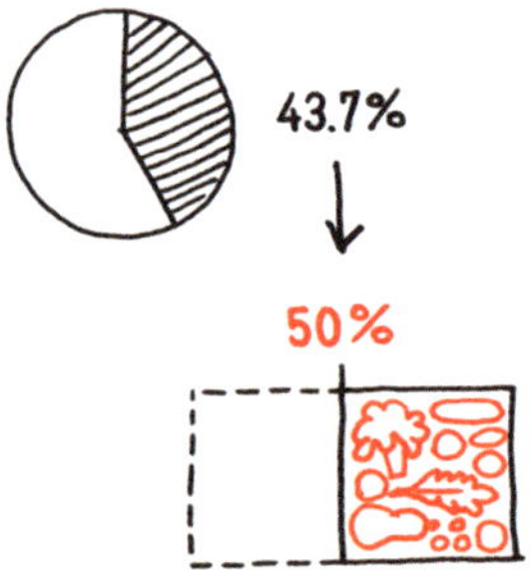

43.7% of community gardens responding to a 2010 survey are growing more than half of their garden as edibles.

SOURCE: 2009-2010 GrowNYC Community Garden Survey

There are more than 490 community gardens in the city, covering just under 100 acres in total area.[16] Most of these gardens provide space for many different activities, including growing vegetables, growing flowers, or simply gathering for socializing or passive recreation. Roughly 80% of these gardens produce some food.

These gardens typically are run by a core group of volunteers or member volunteers – usually drawn from the surrounding neighborhood – who maintain individual plots or communal growing spaces. Gardens on City-owned land are required to host open hours for the general public.

Most of the city's community gardens are located on formerly vacant lots, which proliferated in many neighborhoods of the city in the 1970s and 1980s. Today, more than half of the existing community gardens have been transferred to the jurisdiction of the Department of Parks & Recreation, and about a quarter are owned and protected by land trusts such as the Brooklyn Queens Land Trust; a small number of community gardens are owned by other government agencies or private owners.

Hands and Hearts Garden, Brooklyn

16 Urban Design Lab, The Potential for Urban Agriculture in NYC, Map 3, p. 33

Community Gardens

- Community Gardens growing food
- Community Gardens
- Community Farms
- Commercial Farms
- NYCHA Gardens
- School Gardens

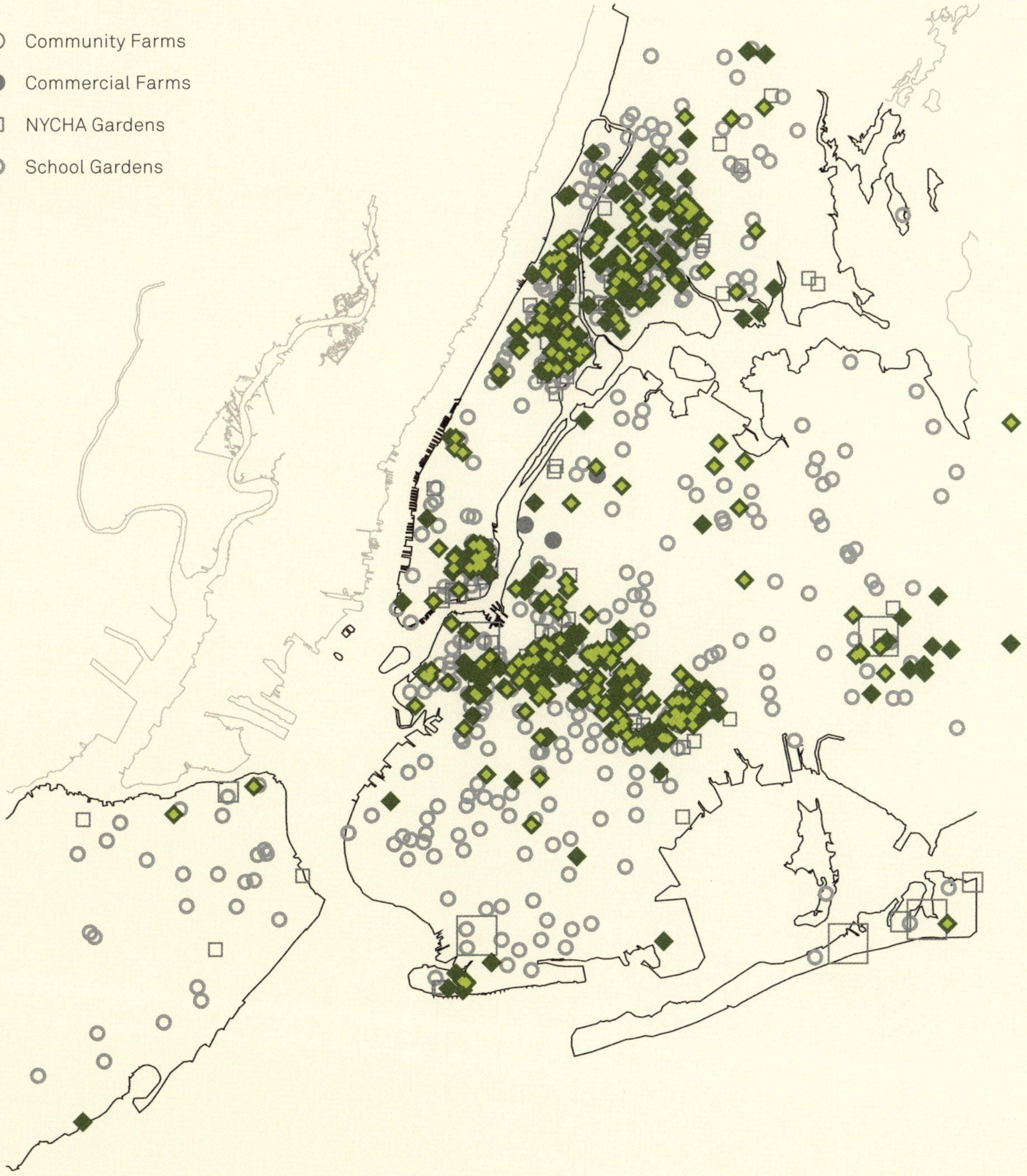

SOURCE: GrowNYC

STAKEHOLDERS

This section briefly introduces three key groups of stakeholders, in addition to farmers and gardeners, who were interviewed for the project:

- **Support Organizations**
- **Government Officials**
- **Funders**

Support Organizations

An extensive network of organizations in New York City provides training, materials, and funding to gardeners and farmers; conducts research and outreach; and encourages elected officials and City agency staff members to develop policies and programs that support urban agriculture.

"[Our work] runs the gamut... doing rainwater harvesting, lending tools to garden projects, we have a plant sale... There is very little we don't do in terms of working with community gardens."

– Support Organization

The work of support organizations falls largely in the following six categories:

1. Provide technical assistance and training for farmers and gardeners

This training includes a broad range of skills and information, from cultivating crops and composting to running an urban farm business. As one support organization staffer described, "Often when there's a community member that says, 'Hey, there's a pest problem, we really need an integrated pest management program in our garden,' we'll send somebody there to do that. And we don't charge for that."

2. Offer funding and resources for programs and site improvements

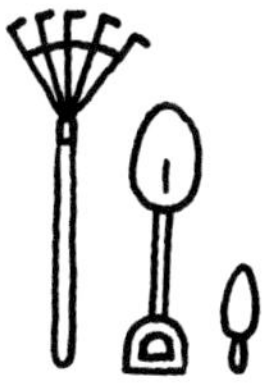

Support organizations provide in-kind items such as plant starts, as well as small grants and loans for materials such as gardening supplies, or for services to refurbish or redesign community gardens. One organization pays a stipend to young people to work at community gardens during the summer.

3. Conduct advocacy and policy work

Support organizations are involved in a broad range of policy work, including advocating for preserving and expanding the land available for urban gardens and farms. During the administration of former Mayor Rudolph Giuliani, for example, many organizations were involved in helping to stop the controversial effort to sell community gardens, and resumed policy work in 2010 during the process of developing new licensing rules for community gardens operating on city property. Intermediaries also have been involved in the effort to legalize beekeeping in New York City, as well as recent food policy efforts led by the Manhattan Borough President and the City Council Speaker.

4. Offer environmental education services

Organizations provide curricula, run classes, and train community members to become educators in the community. As one support organization staffer explained, "We bring high school students to basically experience a community garden and to learn about where their food comes from, environmental connections to food. They also do learn a bit about health and nutrition."

5. Advocate and facilitate systems to increase the quantity of food grown, marketed, and distributed in the city

One support organization trains farmers and gardeners to run farmers markets and assists with outreach and obtaining market insurance.

6. Organize networking and outreach events for urban farmers and gardeners

One support organization recently started a citywide MeetUp group for people interested in raising chickens.

Government Officials

Officials at numerous federal, state, and local government agencies are involved in making urban agriculture possible in New York City, even if individual departments, or programs within departments, do not include urban agriculture as an explicit part of their mission. Government agencies provide equipment and supplies, from lumber to compost; contract with urban agriculture organizations that provide programs and technical assistance; or directly offer technical assistance, logistical support, and construction and maintenance help.

Federal agencies

- The United States Department of Agriculture funds urban agriculture research and program development.[17]
- The Department of Housing and Urban Development funds the GreenThumb program, a division of the NYC Department of Parks & Recreation, through its disbursement of Community Development Block Grants.
- The Environmental Protection Agency provides programs and technical assistance to transform land with contaminated soils into safe sites for growing food.[18]

New York State agencies

- The Department of Agriculture and Markets works to grow the state's food and agriculture industry.[19] The agency supports programs to assist community gardens, enable low-income New Yorkers to purchase food from farmers markets, increase market demand for New York State food, and build the infrastructure needed by agricultural producers throughout the state. Its district office in Brooklyn supports the development of urban agriculture.
- Cornell Cooperative Extension, funded through a federal, state, and local government partnership, provides training in horticulture and ecology.[20]
- Department of Environmental Conservation and the Office of Parks, Recreation and Historic Preservation have provided funds for urban agriculture.

17 http://kyf.blogs.usda.gov/files/2011/10/USDA_Urban_Ag_Memo-Final.pdf

18 http://www.epa.gov/brownfields/urbanag/

19 http://www.agriculture.ny.gov/TheDepartment.html

20 http://nyc.cce.cornell.edu/UrbanEnvironment/Pages/default.aspx

New York City agencies

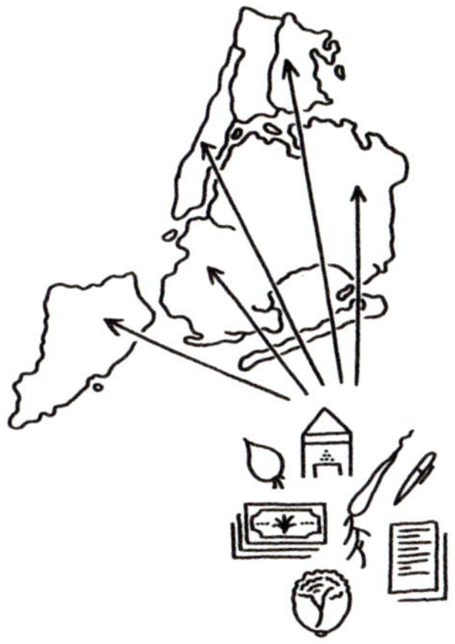

- The Department of Parks & Recreation runs GreenThumb, a program that licenses the community gardens located on city property, offers technical assistance, supplies a number of different material resources, and provides labor to help clear vacant lots for new gardens.[21]

- The New York City Housing Authority has a program manager and several staff who focus on supporting an extensive community gardening program with some 600 gardens on Housing Authority sites, including 245 that produce food.[22]

- The Department of Education hosts an estimated 289 school gardens on school property. The department's SchoolFood division, in cooperation with the nonprofit organization GrowNYC, supports the Garden to School Café program, which uses produce grown in school gardens in school lunches.

- Several other city agencies, including the Departments of Housing Preservation and Development, Education, Transportation, and Corrections control city property on which individuals are growing food.

- The Office of Long Term Planning and Sustainability is involved in strategic planning that includes open space and food production.

- The Department of Citywide Administrative Services (DCAS) is responsible for all city-owned properties. Local Law 48 requires DCAS to publish a list of all city-owned properties that are suitable for urban agriculture.[23]

- The Food Policy Coordinator is responsible for improving food quality served by agencies and access to healthy food throughout New York City.[24]

- The Department of Health and Mental Hygiene is responsible for increasing the availability of fresh fruits and vegetables in underserved neighborhoods and encouraging New Yorkers to increase their consumption of fruits and vegetables. Health Department programs related to urban agriculture include:
    - Health Bucks: coupons that can be redeemed at farmers markets by recipients of the federal Supplemental Nutrition Assistance Program (SNAP)
    - Stellar Farmers Markets: a program to provide nutrition education and cooking classes at select farmers markets

- The Department of Sanitation (DSNY) funds community-based composting programs and has offered leaf and yard waste compost to gardens and farms.[25] DSNY also provides assistance clearing vacant lots that are being turned into community gardens.

21 http://www.greenthumbnyc.org/about.html

22 http://greennycha.org/resources/gardening/

23 Local Law 48 of 2011.

24 http://www.nyc.gov/html/ceo/html/programs/food_policy.shtml

25 http://www.nyc.gov/html/nycwasteless/html/compost/compostproj.shtml Accessed 2/2/2012

- The Department of Environmental Protection, which runs the city's water and sewer infrastructure, has adopted a green infrastructure plan that provides financial support to increase permeable surfaces, including community and rooftop gardens, that stem the flow of stormwater into the city's sewage system, including community and rooftop gardens.[26]

- The Department of City Planning addresses the land use rules that govern where urban gardens, farms, markets, and other facilities can be located.

- The Department of Buildings issues building permits for installations such as green roofs.

- The New York City School Construction Authority, which builds and renovates schools, has been involved in building greenhouses and other growing facilities on school property.

New York City elected officials

Council Members and the Mayor are involved in enacting and carrying out urban agriculture policies, such as the initiatives outlined in the City Council Speaker's FoodWorks 2010 policy plan, including the specific local laws and resolutions that put those proposed initiatives into practice. The Borough Presidents and City Council members also have discretionary funds that they use to provide operating funds to urban agriculture programs and capital funds for garden and farm infrastructure. Finally, the City Council has oversight responsibilities to ensure that programs are carried out effectively and that proposed program budgets are adequate.

26 http://www.nyc.gov/html/dep/html/stormwater/nyc_green_infrastructure_plan.shtml

Funders

Numerous foundations fund urban farm, garden, and support organizations in New York City, and interest in urban agriculture is growing within the philanthropic community. The funders interviewed for this project provide the following support to urban agriculture:

Technical assistance and funding for specific programs

One funder provides assistance to help sustain farmers markets, including handling and processing purchases made with federal food benefits such as the Supplemental Nutrition Assistance Program (SNAP). This foundation also develops and funds innovative programs to provide several different financial incentives that encourage low-income individuals to purchase food from farmers markets.

Capacity building

This includes professional development, training, and board development, or for development consultants and grant writers who can help raise funds.

"We want to grow with an organization, we want to invest in them over time, we want to help them build their internal capacity so that again they are invested in the community's improvement, they're invested in uplifting the entire community."

– Funder

Serving as a liaison to other funders

For example, one funder described a partnership with the W. K. Kellogg Foundation that enabled it to serve as a channel to help disburse Kellogg funds to grassroots organizations that would find it difficult otherwise to obtain a Kellogg grant.

NEEDS & CHALLENGES

Urban gardens and farms require many different resources, including the following:

- Physical components for food production, such as growing space, tools, construction material, soil and compost, water fertilizers, livestock feed and infrastructure (e.g., fences, sheds to store tools, bins to compost food waste)

- Nonphysical components, such as operating funds, labor, permits to run farmers markets or to build rooftop greenhouses, and support services, such as training on best practices for growing vegetables, or opportunities to network with and learn from other farmers and gardeners

This section summarizes challenges farmers and gardeners interviewed for the project cited in obtaining six key resources, including growing space, funding, labor, soil, compost, and construction materials or infrastructure. This section also describes comments about opportunities and challenges related to farmers' and gardeners' interactions with other stakeholders, including: ongoing community engagement, networking with other farmers and gardeners, interaction with government agencies, involvement with policy making, and race- and class-based disparities within the city's urban agriculture movement.

Growing Space

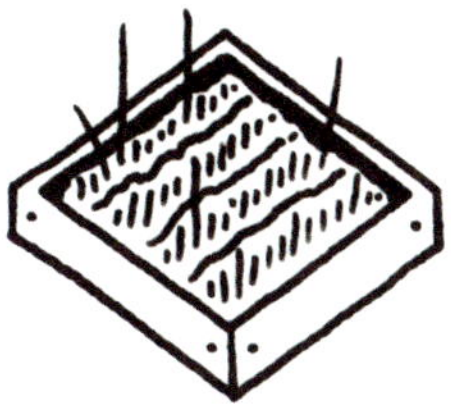

In a densely developed city with high land values, identifying available growing space for urban agriculture and deciding the appropriate tenure for its use has been extremely challenging. Farmers and gardeners interviewed for the project report that the lack of accessible growing space is one of the most significant factors limiting the growth of urban agriculture. Nearly half mentioned land or space as a limit on their capacity to produce food and, in some cases, their ability to expand targeted programs for specific populations.

There are a number of dimensions to this constraint. Farmers and gardeners mentioned the difficulty identifying and gaining access to new suitable sites, either on an interim, long-term, or permanent basis. Other farmers and gardeners mentioned the challenges of growing in small spaces. One commercial farmer noted, "Some of the deep-feeding and long-lasting brassicas, like brussels sprouts, big cabbage heads, and stuff like that, don't tend to be as successful or as worthwhile when you're dealing with a constrained space." Within community gardens, members sometimes compete for space for different activities (growing food, growing flowers, passive recreation, etc.).

Identifying growing spaces suitable for urban agriculture

Rooftops

There is no formal process to link building owners to farmers interested in growing food on rooftops or organizations that run farms. Rooftop farmers interviewed for the project reported that they simply approached individual property owners until they found someone willing to host a project.

Land

Typically, individuals or community groups who want to claim a vacant site for urban agriculture research its ownership and, if it is City-owned property, attempt to contact the City (often GreenThumb) to identify which agency is responsible for the property and determine whether the agency will allow them to use the site for urban agriculture. Government officials are currently developing criteria to assess what City-owned parcels—including vacant lots, underused land, and rooftops—are suitable for urban agriculture.

Land tenure

"I think you're probably hearing a broken record as you talk about this around town—but for us, what we hear the most fear about is tenure, the protection of the gardens."

– Support Organization

On September 13, 2010, the Department of Parks & Recreation published new rules in the City Record governing the licensing of community gardens located on city property.[27] The rules enable a community gardening group to remain in place for as long as the City wishes the site to be a community garden, with an automatic renewal of a garden's license provided that the garden complies with the terms and conditions of its license and meets the registration criteria of GreenThumb.[28] However, if the gardeners fail to comply with those terms or abandon the garden, constituting a "default," the department is required to attempt to identify a successor gardening group and offer it a license for the garden.[29] This involves contacting the local community board and City Council member, and may include contacting nearby gardening groups.

If the City wishes to remove a community garden for a different use of the land, the rules provide for a "garden review process" (see sidebar) that some officials believe makes it too onerous for the City to actually develop these sites. However, some interviewees expressed lingering wariness about the new garden rules, as well as a desire for more assurance that gardens would be protected.[30] "People would love to feel comfortable that they're able to farm and grow food without thinking that ... next year their garden or their land will be lost," said one community gardener. "That overshadows a lot of things. That's the number one [issue]." Government officials interviewed stated that the new licensing procedures effectively make it unlikely that the gardens will be developed. Perhaps as a legacy of the Giuliani administration's development of community garden sites, some gardeners remain worried that they can be evicted even if doing so requires the City to go through a somewhat complex, public process.

27 The City Record, Volume CXXXVII Number 176. Monday, September 13, 2010. pg. 2549. Addition of New Chapter 6 to Title 56 of the Official Compilation of Rules of the City of New York.

28 Compilation of The Rules of the City of New York. Title 56. Department of Parks & Recreation. Chapter 6-03

29 The licensing process also allows the Parks Department to establish requirements with which the gardeners must comply, in addition to applicable federal, state, and local laws, rules, regulations, codes, and ordinances. Failure to comply with any of these also constitutes default.

30 The interviews took place relatively soon after new site licensing regulations for community gardens were developed in 2010, and the concerns expressed by the gardeners interviewed may reflect issues raised during the negotiations over those regulations.

GARDEN REVIEW PROCESS

If the City wishes to remove a community garden for a different use of the land, the rules provide for a "garden review process."[31] The first step in this process involves the Parks Department notifying the garden contact person and alternate garden contact person of the proposed transfer. Parks then provides a list of alternative sites within a half-mile of the garden, or if none exist within a half-mile, sites within the same community board. The department may also state that there is no available City-owned vacant land within a half-mile or the community district, in which case the City is not obligated to provide an alternative site. After being offered an alternative site, the garden has 45 days to either select or reject it. Alternate sites are offered as-is, meaning that any preparatory work must be done by the gardening group. The department's obligation to help the gardening group to relocate is "contingent upon staff availability and resources," meaning that if resources are low the department need not assist with relocation.

Before any transfer of land use occurs, the Parks Department is required to prepare a Garden Review Statement that includes information about the garden and its programming, the alternate site list, and a statement of whether the Garden Contact Person has accepted or rejected an alternative site. At least 45 days before any transfer, the Department must send the Garden Review Statement to the garden contact person, the alternate garden contact person, the City Council member for the council district in which the garden is located, and the Community Board of the community district in which the garden is located. The department is required to include a Garden Review Statement in any application pursuant to the Uniform Land Use Review Procedure (a standardized procedure for the review of all applications affecting land use in NYC by Community Boards, the Borough President, and City Planning Commission), urban renewal and urban development action area programs, the private housing finance code, and the process by which land is leased to local development corporations.

If, as a result of the applicable public review process, the transfer is disapproved, the garden may remain on the site until approval of a future transfer of the parcel. Any new approval of a future transfer also must follow the garden review process. Once the transfer of the garden is approved the Parks Department may terminate a garden's license and order the gardeners to vacate the lot.

31 Compilation of The Rules of the City of New York. Title 56. Department of Parks and Recreation. Chapter 6-05

One mechanism to protect community gardens and urban farms is to turn them over to a land trust or nonprofit organization. Land trusts are membership organizations in which the gardens are the members, and garden representatives act on behalf of their gardens in an official capacity either on the board or as voting members. For example, in 1999, the Trust for Public Land (TPL) purchased 62 community gardens that then-Mayor Giuliani had moved to evict. TPL also received a donation of five additional gardens from the City, and purchased two more gardens for a total of 69 gardens that comprise approximately 8 acres. TPL then helped create three new nonprofit organizations to ultimately own and manage most of these gardens: the Bronx, Manhattan, and Brooklyn-Queens Land Trusts. As of January 2012, TPL has transferred the title to 64 of the 69 gardens to these organizations.

Unlike the GreenThumb gardens, the land trusts have titles to their gardens, which are permanently protected from development.[32] The land trusts enter into one-year renewable license agreements with garden groups that set forth the groups' stewardship requirements. Failure to maintain the gardens can be cause for a license to be revoked by the land trust, as is the case with a GreenThumb garden license.

Financial Resources

Farmers and gardeners reported having a number of different revenue sources, including:

- Sales of produce (through markets and CSAs, to restaurants, etc.)
- Foundation grants
- Government funds
- Proceeds from fundraisers
- Individual donations
- Fees for service [e.g., offering education programs or spaces for farmers selling at the farm or garden's market]
- Fees for visitor groups [e.g., hosting a team-building volunteer day for corporate employees]

The budgets of individual farms and gardens vary significantly. For example, an institutional farmer was able to list numerous sources of funding for infrastructure totaling over a million dollars. On the other hand, a community gardener had hosted a small

32 Land Trust deeds state that the land "shall be used in perpetuity for open space purposes, and for no other purpose." Communication with Joanne Morse, Trust for Public Land, February 1, 2012.

fundraising party that netted $250 in donations for a $550 generator for the garden, with the gardener paying for the rest out of pocket. Most urban agriculture operations have limited financial resources that constrain their ability to grow food, offer programs, hire staff, and make improvements to their farms and gardens. One community farmer explained, "Even if we had more space, if we didn't have another staff member it'd be hard to raise more food. And that's obviously directly linked to funding."

Just over half of the farmers and gardeners in this study reported having a balanced budget, with revenue derived from a combination of sales and grants. One community farmer noted that in most years "revenue was outpacing expenses so we were actually generating a surplus of cash, which we used to build out this physical infrastructure and hire staff...." In contrast, an institutional farmer noted: "We have not been able to cover our salaries, so I would say that we are not yet breaking even. Because we love the project, and because we see a future in our salary, we've been willing to take a big hit this year for ourselves." Additionally, two gardeners reported that their operations could break even if the main coordinator provided pro bono labor, or if their operation focused solely on increasing sales. Several farmers and gardeners said they would like to learn more about business and management models for farming, in order to stretch their limited resources.

"I need to be able to hire....a chief of operations. Or like somebody who has a business degree.... I think in New York City particularly the urban ag movement is populated and run by not-for-profits. We have a different bottom line, and I think that bottom line is important. And I think we need some help from people who understand distribution, understand sales, understand aggregation and to help not-for-profits do that."

– Community Farmer

Labor

More than half of the farmers and gardeners interviewed said that labor was one of their largest costs or challenges. Community gardeners and farmers often struggle to pay staff and offer stipends to interns. One community gardener expressed frustration about the difficulties in paying other community members to run the garden's farmers market while not being able to receive a salary for managing and coordinating the market. Even commercial farmers rely to a large extent on volunteers and unpaid interns for critical tasks. One commercial farmer noted that he was the only one on the farm who received a salary, but that it would be advantageous to be able "to pay someone with some pretty solid experience" during the summer "when there's just not enough hours in the day," to ensure that the work is done more reliably.

In addition to the expense of labor, farmers and gardeners mentioned that finding employees and volunteers with farming and other necessary skills (such as carpentry) was a significant challenge.

Soil

"Soil is probably the number one [production constraint]... [Because of this], we haven't even come close to the number of raised beds that our land could hold."

– Community Farmer

A new 4-by-8-foot community garden bed requires approximately one cubic yard of clean soil—an amount that fills roughly nine milk crates.[33] Existing garden beds must be top-dressed with 2–4 inches of soil (or compost) each year to maintain volume and fertility over time. Soil is expensive to purchase and transport—according to one community farmer, even a poorly performing soil costs about $20/cubic yard, plus another $20/cubic yard to deliver.

Lack of quality soil was a production limitation for nearly a third of the farmers and gardeners interviewed. This was the case for container gardening on rooftops, which requires a lighter-weight growing medium, as well as for some in-ground gardens. One community farmer said, "Everybody needs soil, [which is] probably the hardest thing to get."

33 John Ameroso; Ursula Chance; personal communication, October 3, 2011.

Compost and Nutrients

Both erosion and agriculture draw nutrients from soil, which farmers and gardeners must replenish in order to continue growing crops. Soil can be replenished in a couple of ways, including the addition of compost and other soil amendments.

Access to sufficient quantities of compost and single-source nutrients (e.g. fertilizers) was another important limiting factor, particularly for new sites with poor soil quality and for rooftop farms, which typically cannot use conventional topsoil because it is too heavy. One commercial farmer noted, "All of those [soil amendments] are cost prohibitive for us at this point until we find a [less expensive] alternative or a donor. It doesn't mean things won't grow but it does mean that ... I've been thinking a lot about ... how nutritious the food is that I grow." An institutional farmer added, "We want a lot of compost [and] it becomes a problem because there is no place to get it for a reasonable price."

"The fact that the municipal leaf composting project got scrapped [has been] a huge detriment to the gardens. That's one policy that would be nice to see reversed."

– Support Organization

Materials and Physical infrastructure

In addition to raised beds, farmers and gardeners may build any number of structures on-site to support their operations: sheds and coops to help protect tools, materials, and livestock from the elements and from theft; hoop houses and greenhouses that enclose vegetables, allowing farmers and gardeners to extend the growing season into the winter; compost bins and rainwater barrels to supplement their soil and water sources.

Several farmers and gardeners said that site security and maintenance were important concerns. In some cases security and maintenance problems actually limited production capacity. Theft was a problem at a few sites, and several operations that were located on City-owned land had experienced difficulty in getting fences installed to protect both the garden supplies and produce, as well as to provide personal safety to gardeners and farmers. A community gardener noted a dilemma in terms of the need to secure the physical assets that gardeners had brought to the gardens, and posed the question, "How do I secure my assets [in a garden] when it is a public space? ... I can't put up barbed wire because we are on Parks' property..." Another gardener described the problem of theft: "The more you get valuable food in, the more people try to get in and move it out [i.e., steal it]." Yet another gardener recounted feeling personally vulnerable while working in the community garden, due to the neighborhood surroundings and lack of adequate fencing around the site.

Farmers and gardeners described a number of ways in which they secured infrastructural improvements: asking building owners to pay for building renovations to start a rooftop farm; securing grants for infrastructure like greenhouses; and obtaining donations of material to build other infrastructure. Farmers and gardeners also source free or low-cost materials such as production supplies and lumber. A commercial farmer who had scavenged crates, bins, and a refrigerator for the farm described keeping costs manageable by "just being frugal ... constantly, and [by being] resourceful."

Ongoing Community Engagement

Overall, the farmers and gardeners interviewed reported that members of the communities surrounding their farm or garden—including local residents, neighborhood organizations, schools, and local business owners—were very supportive. One community farmer said, "I can walk five blocks in either direction and [people] will recognize our work." Neighborhood residents also show their support by participating in farm and garden activities: volunteering, attending public events, fundraisers, and farmers markets, joining community-supported agriculture (CSA) programs or campaigns to preserve community gardens. In addition to support from individuals, local businesses and neighborhood organizations also support urban agriculture operations through in-kind donations and by allowing farmers and gardeners to use their meeting spaces.

Yet farmers and gardeners noted that sustaining this neighborhood connection and support was sometimes challenging. In the case of community gardens located on city property, avoiding losing a critical mass of active members is necessary to renew the licenses that allow the garden to operate. One community gardener noted the importance of "getting more public awareness, really getting the community at large involved in the work that we do."

"It's time to stop speaking to the choir and going out into the community and having town hall meetings and going to churches or rec centers or whatever and get this

message out to the people so that they can become stakeholders [in the garden] and not bystanders."

– Community Gardener

These efforts are not confined to community gardeners. One commercial farmer remarked, "We are trying to build [a stronger] neighborhood network because ... the whole point is to get your base to be your biggest supporters." Some farmers and gardeners described trying to strengthen or expand their relationships with specific populations, notably low-income neighborhood residents, native Spanish speakers, and youth.

Networking with the Urban Agriculture Community

The city's farmers and gardeners work with each other, and with support organizations and government agencies, in many ways. Farmers and gardeners share resources (such as supplies or tools) with each other or provide direct assistance—one community farmer, for instance, described how the farm helped the president of a nearby garden raise funds. Farmers and gardeners also work together on policy-making. A community farmer remarked that farmers meet at the Bronx Borough President's Food and Sustainability Coalition to "bring visibility to the efficacy of urban farming" and ... [encourage] the Borough President to buy into that vision."

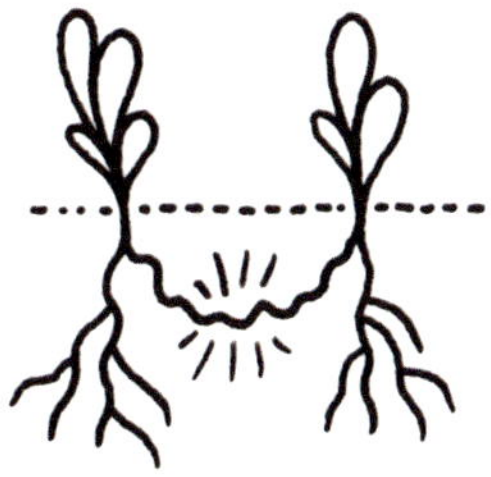

These partnerships extend beyond city limits. Urban gardens and farms often do not produce enough volume or variety to support a full-fledged farmers market, and certain market assistance programs require that markets sell food from more than one farm. Partnerships with regional farms help make urban farmers markets viable.[34]

Several interviewees discussed wanting to strengthen their relationships with farmers outside of the city and to forge new connections. For some, the goal was to expand their production capacity through partnerships with larger landholders. Others sought business and growing advice from farmers whose focus was on production and sales rather than the more typical nonprofit business models found in cities.

34 New York State Department of Agriculture and Markets, 2011 New York State Farmers' Market Nutrition Program (FMNP). Participation Requirements for Farmers' Markets. Rev. 3/2011. Accessible at http://www.agriculture.ny.gov/AP/agservices/fmnp/H-2011_FMNP_MarketRequirements.pdf

"It is invaluable for me to go and sit down with a farmer, somebody that's been doing it for thirty years... and I don't really have that [opportunity] in the city. People don't really come from the same commercial concept, you know, it's a little bit more towards the nonprofit status, or making all their money [hosting] children's groups, or whatever. So I don't have that parallel anywhere in the city, I have to go to Long Island or upstate to find that."

– Commercial Farmer

One community gardener has discussed with a farmer who sells at his farm stand the possibility of creating a program in which young people from the neighborhood would spend time upstate during part of the growing season learning about farming and experiencing life in a rural community.

This networking extends to the philanthropic, policy, and business communities. Several farmers and gardeners said that they hoped to cultivate relationships with funders, state agencies, and local restaurants.

Interactions with Government Agencies

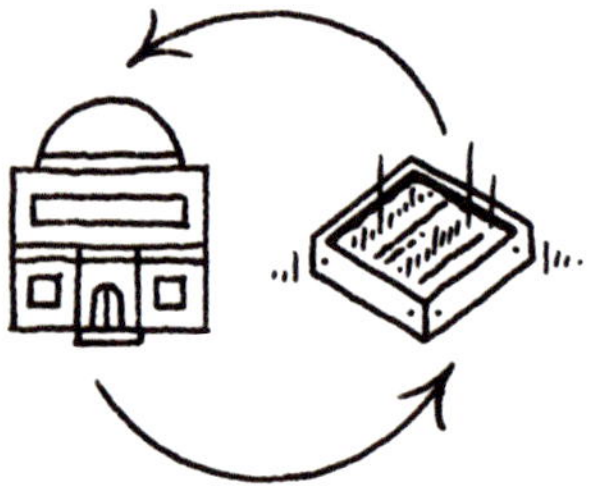

The urban farmers and gardeners interviewed reported having good relationships with government agencies that provided support for their agricultural activities. However, they also offered examples of how agency rules and regulations were unclear or even stymied their work. And there was a general desire among interviewees for agencies to do more to support and help expand urban agriculture activities.

Several interviewees mentioned how program cutbacks made it difficult for gardens and farms to pay for materials and services that had been provided by an agency. In particular, farmers and gardeners described the difficulty in finding substitute sources of compost when the Department of Sanitation's Compost Giveback Program was eliminated. Others mentioned budget cuts to services such as the Department of Youth and Community Development's Summer Youth Employment Program and MTA's Access-a-Ride program, which had facilitated participation in urban agriculture activities by youth and seniors, respectively.

Other farmers and gardeners expressed frustration with delays on the part of agencies, confusing procedures, and what they considered to be onerous regulations. For example, one community farmer mentioned a struggle to obtain farmers market coupons for seniors in time for the start of the growing season: "They don't come out, or [the agency] stalls on sending them out until August, so for a lot of seniors it means that they don't have access to fresh food until [then]." Others noted that they faced difficulties obtaining permits for community farmers markets and community kitchens, and that permit requirements were unclear.[35] For example, a commercial farmer who built a chicken coop described facing a fine because the coop was considered an additional structure on the property. A community gardener described the problem of obtaining a permit for a commercial kitchen in a community center so that the gardeners could use it to produce value-added foods: "We were this close to certification and the Fire Department came in and said that to have a certified kitchen you have to have some sort of special [fume] hood... and the hood cost $30,000." Another community gardener urged "clearer guidelines for farmers markets and for selling produce."

"The permit process is really convoluted and ever-changing. Sometimes you don't need one at all. [At one of our markets] because the market wraps around a corner we have to get two permits, so we have to pay double what most markets would. So that can be frustrating."

– Community Gardener

Some farmers and gardeners expressed a need for greater responsiveness on the part of some City agencies related to site maintenance, sanitation, safety, and security at gardens, farms, and farmers markets located on public land. For example, one community gardener described the complex process of getting the City to remove a squatter living in the garden that required the involvement of several different agencies. "[In the end, one agency] came and took the tent down, and I don't know if... we finally just talked to the right person [but]... I don't know why it took so long or why [it finally was resolved]."

35 For more information on the challenges of operating community farmers markets in New York see Red Tape, Green Vegetables: A Plan to Improve New York City's Regulations for Community-Based Farmers Markets, April 2011.

Involvement in Policy-making

A number of farmers and gardeners interviewed wanted to be more connected to the policy-making process, claiming that community members were often excluded from this process or included only nominally in policy decisions affecting their neighborhoods. Some farmers and gardeners felt that City agencies did not sufficiently appreciate the benefits of the grassroots community organizing related to food, agriculture, community safety, or environmental remediation that their gardens and farms were involved in, and as a result failed to take advantage of their expertise and networks to implement agency programs. For example, one community gardener expressed the opinion that the Department of Health had not sufficiently considered the value of community gardens in its health work. Moreover, the gardener felt that the agency had taken more than it had returned to the urban agriculture program, engaging gardeners to glean information from them rather than to better support their activities as partners. "I'm not going to say that there's no value to that relationship [with the Health Department], but I believe that relationship is about absorbing instead of supporting."

This feeling of disconnection with respect to policy-making was also expressed as a sense that government officials make decisions based on citywide criteria, and in doing so they were neglecting critical block- and neighborhood-level concerns. For example, one community farmer criticized the community garden license provisions that require the Parks Department to offer alternative sites that are within a half-mile of any City-owned garden proposed for development, or a site within the community board if no appropriate vacant land is within a half-mile. The community farmer argued that urban communities are "super-hyper-local," and that residents might feel no connection to alternative garden sites a half-mile away. "There is no understanding because they are just using numbers," the farmer stated.

"The issue we have at [my department] and at all City agencies is that we are trying to hit millions of people. It's great if a garden can produce enough food to feed ten people, but that's not something where we're going to work on a policy or program around because we're truly trying to hit as many people as possible."

– Government Official

Several interviewees said that the perception that urban agriculture has not yet reached a significant scale influences government officials' willingness to support urban agriculture at a policy level. As one government official explained:

> The way the Mayor's people are looking at [urban agriculture], except for the ones who are specifically assigned to promote it, they're just going to be like, 'What is that? It's a drop in the bucket. How much more tax revenue are we going to get off of some new [development] that [could] go there?'... I just don't ever see it being a real priority."

Race- and Class-based Disparities within the City's Urban Agriculture System

Urban agriculture contributes to the economic, social, health, and ecological improvement of communities. These benefits are particularly important in underserved neighborhoods, including many low-income communities and communities of color.

However, a number of interviewees, including farmers and gardeners, support organization staff, and funders, suggested that significant race- and class-based disparities exist within New York City's urban agriculture system, producing negative impacts for individuals, communities, and the system as a whole. Some farmers and gardeners cited inequities in access to philanthropic dollars, government grants, and in-kind assistance, as well as to information about these opportunities. There is also wide variation in farmers' and gardeners' capacities to write competitive funding proposals and to evaluate the impacts of their programs. The consequences of these disparities can be significant, with some farms and gardens able to gain access to hundreds of thousands of dollars in city and state funding, and others resorting to in-kind donations from individuals and grassroots fundraisers to afford the most basic supplies and repairs.

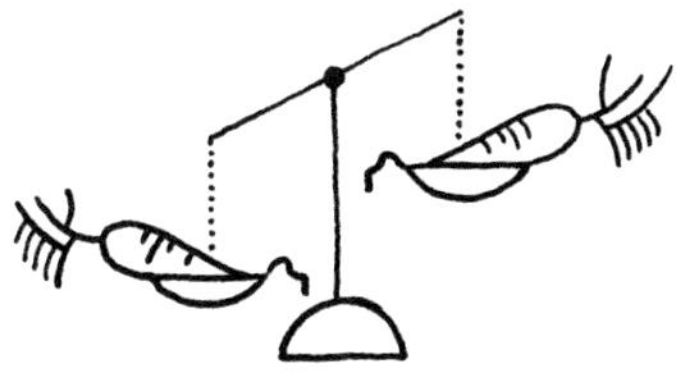

Several interviewees characterized New York City's urban agriculture sector as two distinct communities, one with significantly more financial resources, stronger relationships with influential groups, and/or a white leadership that has created or taken advantage of opportunities to expand their operations. These opportunities have been unavailable to or more difficult to attain by a community of minority-led, less connected organizations struggling to raise funds to cover their basic operating costs. A community gardener cautioned against allowing the urban agriculture movement to remain divided into "the haves and the have-nots."

"I'm afraid right now that the way [urban agriculture is] looking is White-led. And that people of color are being pushed to the side. I don't want crumbs... And I want to make sure that if this movement is sustainable that it is has to be equal, because right now I'm starting to see a trend whereby the people with the most power, the most voices, are getting the money and the people who can't speak as well are [not]."

– Community Gardener

Several interviewees were confident that urban agriculture could be a mechanism for political and social change to reduce race- and class-based disparities, provided that all farmers and gardeners are able to have a voice in policy-making. One community farmer said that food "can empower people to have political...and economic power," which is part of the mission of [our] farm." This farmer expressed an interest in helping people of color in urban agriculture to become more involved in different segments of the urban agriculture system, such as in farmers markets, "so they don't get left out of a growing marketplace. And I don't mind them growing food just to give it away, which they have been doing, but they have to have a voice." An institutional farmer noted further that

> There are two very unique and distinct aspects of this urban farm movement that are going on.... One is very middle class and White, and one is not. One is of-color and very low-income. And they are ... very separate. Unless they are brought together, I don't know that the success of either is going to continue. The needs [of each group] are completely different.

As urban agriculture continues to grow in popularity, and competition among gardeners and farmers for government and foundation funds grows, these disparities are likely to become more acute, not less, unless steps are taken to overcome systemic inequities.

Conclusion

Long time observers of the city's urban agriculture movement have noted that they are no longer able to keep track of the constant emergence of new farms and gardens citywide. At the same time, a growing number of funders, journalists, elected officials, business leaders, and other New Yorkers are learning about how urban agriculture transforms vacant land and empty rooftops into community resources, and helps government agencies achieve many pressing municipal goals.

The remaining chapters of this publication address two critical, interrelated opportunities to support and expand urban agriculture in New York City. The Metrics chapter establishes a framework to allow New Yorkers to understand how the urban agricultural activities taking place at the city's farms and gardens contribute to a wide range of benefits, and provides tools for farmers and gardeners to track the scope, scale, and impacts of their work. The Policy chapter provides a roadmap to help all key stakeholders – from farmers and gardeners to government officials to volunteers – to coordinate their actions, expertise, and resources.

Garden of Happiness, Bronx

II. METRICS

INTRODUCTION

Many of the government officials, support organizations, funders, and farmers and gardeners interviewed for *Five Borough Farm* wanted to understand how urban agriculture contributes to a wide range of outcomes, from people's eating and exercise habits to their social connections with their neighbors, and from job training and readiness to environmental factors such as air quality and stormwater retention. Numerous studies have shown that urban agriculture can contribute to social, health, economic, and ecological benefits. However, very little of that work focuses specifically on New York City, or develops measures that can demonstrate benefits at a neighborhood as well as a citywide scale, and ultimately inform the decision-making of elected officials and agency commissioners.

How can government officials—or anyone interested in urban agriculture, for that matter—find out how much food is produced in New York City? For now, the answer is: they can't. There is no official comprehensive count of the number of urban agriculture sites citywide, how much food they produce, how much food waste they divert from the waste stream that the City must otherwise pay to ship to landfills, or how many students participate in farm- and garden-based classes. A few agencies or organizations collect some of this information—the Farming Concrete project, for instance, measured food production in 67 community gardens citywide in 2010. However, no one has yet addressed the full range of urban agricultural activities, or developed a framework to conceptualize, prioritize, and capture the aggregate impacts of the more than 700 urban agriculture sites citywide, or estimated what these impacts might be if urban agricultural activity expanded.

This chapter establishes a metrics framework to understand how the broad range of activities taking place at the city's farms and gardens can contribute to social, health, economic, and ecological outcomes. This framework includes a recommended set of indicators that farmers and gardeners can use to track urban agricultural activities (such as the production of food and compost, employment, sales, and participation in programs), as well as more complex outcomes (such as the attitudes of young people who work at a community garden, or the eating patterns of people who receive food from a community farm). The recommended indicators are designed to make the process of data collection and analysis accessible to anyone in the urban agriculture community, allowing even farmers and gardeners with limited resources to report on their activity, thus making it easier to aggregate information on urban agriculture's impacts citywide.

Data collected using these indicators ultimately could inform citywide urban agriculture policy. For instance, the Department of Environmental Protection (DEP) has committed to invest $187 million over the next four years on stormwater management strategies that include pilot urban agriculture projects on privately-owned property. Urban agriculture sites create permeable surfaces that capture stormwater, and scores of farms and gardens have installed rain barrels to capture and use stormwater for their vegetable beds, diverting it from city sewers. An accurate count of how many sites are involved and how many square feet of permeable surface is represented could help agencies like DEP—as well as funders and support organizations—gauge how urban agriculture contributes to its goals, determine if they want to expand their support for it, and design programs and grants to better meet the needs of urban farmers and gardeners.

Indicators

Indicators are signs of progress and change that result from an activity, project, or program. Indicators can be compared to road signs which tell travellers if they are on the right road, how far they have traveled, and how much distance is left to reach a destination.

FARMING CONCRETE

TRACKING FOOD PRODUCTION

In 2010, the 97th Street Block Association Community Garden in Queens produced an estimated two tons of food on about 1200 square feet of land dedicated to growing food – an area roughly equal to the inside of a New York City subway car. The gardeners and a group of volunteers calculated this number, in part, by weighing vegetables on a garden scale and mapping how much of the garden was under production.

The 97th Street garden was part of an initiative called Farming Concrete, launched in December 2009 to measure food production in a sample of the estimated 400 NYC community gardens that grow food.[36] In 2010, volunteers reached out to community gardeners throughout the city to enlist their support in weighing the produce, mapping their garden beds, and keeping track of the crops they grew, as well as recording personal reflections about their gardening. The 2010 data showed that 67 sampled community gardens, a total of 1.7 acres, grew approximately 87,700 pounds of fresh produce worth more than $200,000. The project has trained interested gardeners to train others in the measurement process, and has also engaged school gardens in tracking their productivity.

Developing a Metrics Framework

The first step in developing a metrics framework was to understand the kinds of benefits that farmers, gardeners, and other key stakeholders intuitively link to urban agriculture. Through interviews and site visits, the team compiled a list of activities that take place at New York City farms and gardens. These include agricultural activities such as growing fruits, vegetables, and herbs, raising bees and chickens, selling produce through farmers markets and Community-Supported Agriculture shares (CSAs), and collecting food waste for compost and rainwater. This list also included educational programs, community organizing campaigns, initiatives directed at specific populations (e.g., women, seniors, youth), and social events. Interviewees cited many benefits they believed resulted from these activities, including greater social interaction, a stronger awareness of and concern for the environment, a greater willingness to engage in social justice issues, and healthier eating patterns.

36 See www.farmingconcrete.org

The next step was to review the extent to which these benefits have been researched and documented through published evidence. The Metrics Fellow surveyed the peer-reviewed literature to identify studies that demonstrated these kinds of benefits. Some highlights from these findings are summarized below:

- Community gardens have been found to improve healthy eating, for instance, by increasing fruit and vegetable consumption and preference among participants, regardless of setting or population.[37] This effect also held true for household members who did not personally garden.[38]

- Studies pointed to many other kinds of health improvements due to participation in urban agriculture, such as a greater amount of physical activity among youth and adults.[39]

- Farm-to-school initiatives that increase availability of fresh produce, reduce prices, and provide point-of-purchase information have been found to be effective strategies to increase fruit and vegetable consumption in schools.[40]

- Studies have found that participating in community gardens built social capital and bonds such as mutual trust and reciprocity, leading to a stronger overall sense of community.[41]

- Community gardens provided wage-earning opportunities for youth, especially for those considered at-risk or from low-income families.[42] In addition to the monetary benefits, youth employment develops job-readiness skills.

- Participating in community gardens has been found to produce improvements in behavior and interpersonal skills, including respectfulness, commitment, and ability to work in teams.[43]

Based on this body of research, *Five Borough Farm* defines nineteen outcomes toward which urban agricultural activities in New York City can contribute. The potential outcomes were grouped into four main categories—health, social, economic, and ecological.

37, 38, 39, 40, 41, 42 and 43 See p.106 for full citations of these studies

METRICS FRAMEWORK

Many studies have shown links between the urban agriculture activities across the top row (such as cooking and nutrition classes, rainwater harvesting, farmers markets) and the health, social, economic, and ecological benefits on the left hand column (such as health eating, stormwater management, and social connections).

◉ Evidence-based links

Benefits	Rainwater Harvesting	Seed Saving	Raising Livestock	Bee Keeping	Composting	Land Remediation	Planting Trees	Environmental Education	Cooking & Nutrition Classes	Food Systems Education	Health & Wellness Education
Health											
Access to healthy food			●								
Food-health literacy		●						●	◉	●	●
Healthy eating									◉		●
Physical activity			●				●				●
Social											
Empowerment & mobilization					●		●	●			
Youth development & education			●						●	◉	●
Food security			●								
Safe spaces			●			●					
Socially integrated aging											
Economic											
Local economic stimulation											
Job growth											
Job readiness								●			
Food affordability			●								
Ecological											
Awareness of food systems ecology		●		●	●			●		●	
Stewardship	●	●			●	●	●	●		●	
Conservation	●	●			●	●					
Storm water management	●										
Soil improvement					●	●					
Biodiversity & habitat improvement			●	◉		◉	◉				

Activities

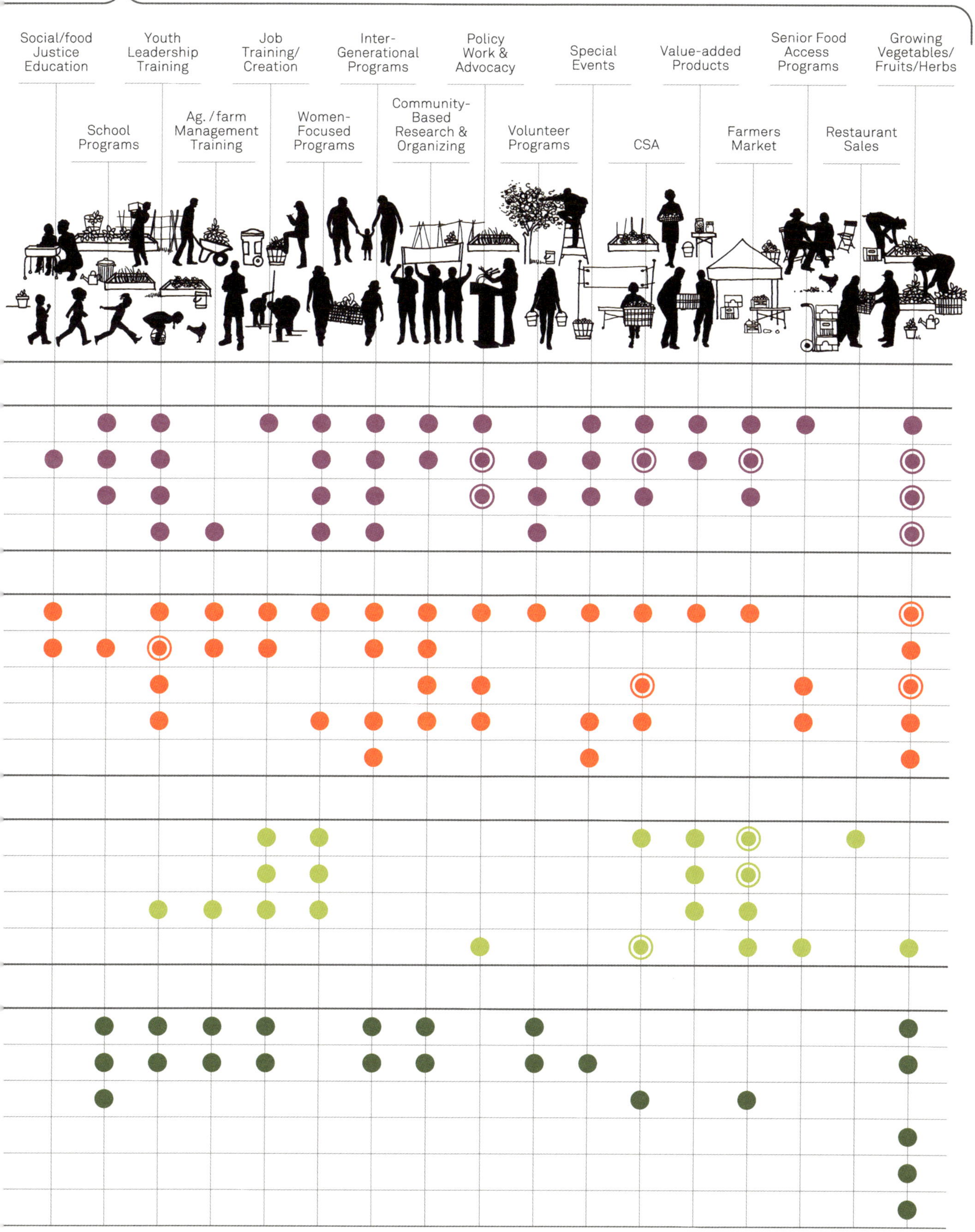

Social/food Justice Education
School Programs
Youth Leadership Training
Ag./farm Management Training
Job Training/ Creation
Women-Focused Programs
Inter-Generational Programs
Community-Based Research & Organizing
Policy Work & Advocacy
Volunteer Programs
Special Events
CSA
Value-added Products
Farmers Market
Senior Food Access Programs
Restaurant Sales
Growing Vegetables/ Fruits/Herbs

Within this framework, activities may contribute to multiple potential benefits. For instance, consider a typical community farmers market in an area with few stores that sell fresh produce. The market employs young people and local residents to manage the market, set up tables and tents, and deliver food. Locals gather at the market to socialize and shop, some using government-sponsored coupons that assist low-income residents in purchasing fresh produce. At the market, staff from nonprofit support organizations provide free classes on health and nutrition, and cooking demonstrations.

Finding published evidence linking programs to outcomes does not mean, for instance, that any youth leadership program or a cooking class automatically changes participant behavior or diet. To demonstrate those kinds of benefits, data must be collected about specific activities that take place at New York City's farms and gardens.

Prioritizing Specific Metrics: Assessing Capacity and Need

What aspects of the broad range of urban agricultural activities should farmers and gardeners track? To answer this question, the Metrics Fellow considered what interviewees said they were interested in tracking, including what government officials said would help them understand urban agriculture's impacts, and assessed the capacity among urban agriculture stakeholders to track farm and garden activities.

Capacity to collect data on urban agricultural impacts

Across the urban agriculture community, interviewees said that lack of staff time, funding, or technical expertise limited their ability to measure the impacts of urban agriculture. One government official remarked that scheduling and conducting site visits to gather data on community gardens and to interview gardeners is extremely time- and labor-intensive. "If somebody's there you talk to them," said this official. "Or maybe you [don't have time] to talk to them because you want to get to six more sites. [And] is that person the main person who actually knows? I mean there are lots of problems with that [approach]." To augment their staff, support organizations sometimes hire interns or outside consultants to conduct and analyze surveys and interviews, or rely on students to help with evaluation. "I don't have time to track anything that someone is not asking me for," said one support organization's director. "So unless a grant funder [says], 'We need those numbers,' I am not keeping track of it."

Capacity to track and evaluate activity varies significantly across the city's farms and gardens. For instance, some farmers and gardeners—particularly commercial farmers—keep sophisticated records about planting methods and outcomes, and the most efficient use of labor. Farmers and gardeners also cited a wide range of formal and informal methods for tracking shopper preferences at farmers markets and the outcomes of youth programs. However, a number of farmers and gardeners said they would need training and funding to rigorously collect and analyze data on their programs. One funder provides grants to support farmers and gardeners evaluating their programs, noting "the vast majority will [require assistance].... Even those [organizations] with the higher level of capacity will ... be stretched."

> *"We're at the stage of developing [an evaluation] process... A lot of donors ask us how we are going to evaluate, and we describe what we think we would do, but then we don't get the grants, we don't get the sponsors."*
>
> – Institutional Farmer

Potential for urban agriculture metrics to influence policy

Government officials interviewed cited specific kinds of information that would help them evaluate the impacts of urban agriculture. Food production was one key indicator. One government official said, "To actually measure urban food production would be amazing, because it's great to keep saying we've got this many gardens and this many farms, but [production figures would help us] to understand the impact." Another official believed that urban food production has the potential to create jobs, and wanted to measure that production before committing additional land. "We want to know how much produce those farms are going to be producing," this official said. "I want to know in term of pounds. I want to know compared to acreage."

Interviewees also wanted to know not only how much food is produced but also how and to whom it is distributed. Speaking about urban farms, one government official said:

> We want to understand who their markets are. Are they at 'x' number of farmers markets? ... Are they selling into stores, and if so, how many stores? [This] gives us a sense of who is benefiting from that farm ... beyond just the farmers getting their revenue.

Another official wanted to demonstrate to state policy makers that urban farms and gardens distribute food in multiple ways to their surrounding neighborhoods, and connect city residents to rural farmers as well:

> I believe that the [New York State] administration views agriculture as a wonderful community greening activity for our urban areas, but there needs to be some education on how important it is to food access, and not just because it grows food, but because it also obviously brings in farmers markets and CSAs, and it brings in other upstate food.

The extent to which urban agriculture programs engage individuals in the surrounding communities can be measured in a number of different ways, including simply by counting the number of people who are served by gardens and farms, either as growers, participants in programming, or customers at farmers markets. "With a school garden we want to know how many kids participate," one official said, "because sometimes it's just one classroom and sometimes it's the whole school." This official continued:

> We know that all these projects are dependent on volunteers, but [it would be useful to have] some kind of idea about the demographics of the volunteers, how [gardens and farms] look for their volunteers... and what happens with the volunteers [over time].

Recommending Urban Agriculture Indicators

"One of the things that people [were] trying to figure out, back when the auction [of NYC community gardens] was happening, [was] what value is this garden?... How can you add up all the volunteer hours that people put in? How can you count up all of the dollars that they might spend to buy garbage bags, or a tool, seeds, whatever? What's the environmental value in terms of biodiversity?"

– Support Organization

Based on the analysis of capacity within the urban agriculture community to track indicators, and the potential of certain indicators to inform policy-making, the Metrics Fellow worked with an advisory group of farmers and gardeners to recommend two types of indicators to track activities and outcomes.

Indicators to track activities

The first type of indicator allows farmers and gardeners to track the scope, scale, and geographic concentration scale of a given activity. For example, how many young people receive structured job training? How many people attended a cooking and nutrition class? How many pounds of food or compost are produced? How many CSA members does a farm or garden have? These indicators can also reveal more fine-grained information. For example, among those young people receiving job training, how many were 15 to 19, and how many were 20 to 24? How many of the CSA members are low-income residents? These can be useful benchmarks for farmers and gardeners to understand the impacts of their programs and changes over time. Because these indicators can be tracked at any number of sites, the data also can be aggregated to demonstrate the cumulative impacts of urban agriculture citywide.

Examples of indicators to track activities:

- Pounds of food produced
- Number of young people participating in a farm/garden program
- Number of people employed by a farm/garden program
- Number and type of activities a program implements related to increasing biodiversity, beekeeping, native planting program, land restoration
- Dollar amount ($) and percentage of sales that are from food access programs

Indicators to track outcomes

The second type of indicator addresses more complex questions of impacts on health, behavior and learning. How many of the people who attended a health-related program report they are eating a fruit or vegetable once a day? Does working at a farm or garden change what local residents feel about safety in their community? These questions cannot be answered simply by tracking participation in programs, but require surveying participants and analyzing the resulting data.

"There isn't a chart of the last ten years of youth work and where everybody has exited, and where they are now. It's a real deficit of the organization, it's a major deficit in the movement."

– Community Farmer

Examples of indicators to track outcomes:

- Number and percentage of people participating in a health/nutrition program who report eating the federal government's recommended five fruits and vegetables per day
- Number and percentage of people participating in a health/nutrition program who can identify where their food comes from (i.e., origin of food as plant-based)
- Number of youth farm/garden participants who graduate from high school
- Farm/garden participants' perceptions of safety (personal safety, crime) in their community

Metrics Implementation

Collecting and analyzing data on urban agriculture takes time, effort, and some technical knowledge; most interviewees, across all categories, said they lacked this capacity. Government agencies, funders, support organizations, and other institutions all should continue to play a role in supporting the collection of data on urban agriculture's impacts. Graduate students who are interested in doing field work on urban agriculture could be enlisted to work with farmers and gardeners to collect data. Agencies and funders could provide grants specifically geared toward facilitating data collection and program evaluation. Ultimately, because urban agriculture is a community-based movement, the collection and tracking of data on urban agricultural activity must include farmers and gardeners, not just outside researchers and government officials. Collecting and analyzing data can be part of a broader educational or community organizing mission, engaging youth and local residents to participate in establishing the value of community gardens and farms. Building farmers' and gardeners' capacity to track and evaluate their activities will help them improve their programs and communicate the benefits of their work to the general public, funders, and policymakers.

Over the next year, the Design Trust will work with a sample of farmers and gardeners citywide to track a few common indicators, including pounds of food and community participation. The companion website to this publication, fiveboroughfarm.org, allows farmers and gardeners to upload data they collect, and download simple reports that show changes over time. The website will also allow users to aggregate data across many different farms and gardens citywide to inform policy.

Five Borough Farm's metrics framework and recommended indicators are the first of their kind to address the full range of urban agriculture activities and benefits. In underserved neighborhoods, where deficit-based metrics typically focus on indicators of poverty, obesity, and truancy, this framework will help people demonstrate how urban agriculture can positively transform communities. Ultimately, this framework could be applied throughout New York City, as well as other municipalities where urban agriculture has taken root.

RECOMMENDED INDICATORS
BY TYPE OF BENEFIT

Health

Urban agriculture can help improve access to, and affordability of, healthy food for underserved communities, improve people's understanding of what makes food healthy or unhealthy, increase healthy eating, and increase physical activity. This in turn contributes to important health outcomes such as lowering rates of obesity, diabetes, cardiovascular disease, hypertension, and stroke.

Improve access to healthy food for underserved communities

Unless otherwise noted, data can be collected through a tracking form.

Indicators marked in color require special survey or study.

(pounds) of food produced by the farm/garden

#($),% of the farmers market sales from food access programs

# of CSAs linked to the farm/garden

# of participants in CSAs linked to the farm/garden

#, % low income shares in CSAs linked to the farm/garden

Improve food-health literacy[44]/skills/aspirations for underserved communities

of people participating in farm/garden programs/activities reporting that they sometimes, very often, or always read food nutrition labels at the supermarket

# of people participating in farm/garden programs/activities that agree with the statement "I can change the things I eat"

44 Health Literacy is defined by the Centers for Disease Control and Prevention as the capacity to obtain, process, and understand basic health information and services to make appropriate health decisions.

Increasing healthy eating

of participants in the farm/garden's health-related programs

# of healthy eating program strategies being implemented

#,% of healthy eating program participants meeting the Healthy People 2020[45] recommended servings of five fruits and vegetables per day

#,% of healthy eating program participants consuming a fruit and/or vegetable snack once per day

# of different vegetables eaten in the last 2 weeks by healthy eating program participants

#,% of healthy eating program participants consuming fast food once a week or less

#,% of healthy eating program participants that can identify where their food comes from (i.e. origin of food as plant based)

Increasing physical activity

of people engaged in farming/gardening on the farm/garden

# of total person-hours spent farming/gardening on the farm/garden per year

(average) time spent farming/gardening among people who participate in programs/activities

45 Healthy People 2020, coordinated by the U.S. Department of Health and Human Services, provides science-based, 10-year national objectives for improving the health of all Americans.

Social

Urban agriculture can promote thriving communities by bringing people of all ages together, developing young peoples' skills and knowledge, and creating safe spaces.

Social capital/connection

Unless otherwise noted, data can be collected through a tracking form.

Indicators marked in color require special survey or study.

- # of people participating in farm/garden programs/activities
- # total person-hours spent working on the farm/garden per year
- # of farmer's markets the farm/garden sells in
- # ($) from urban agriculture-based farmers market sales
- % increase in urban agriculture-based farmers market sales from prior year
- community members' perceived benefits and concerns about urban agriculture
- community perceptions of safety (personal safety, crime) in their community

Youth development

- # of youth participating in the farm/garden
- #,% of farm/garden programs that employ youth
- # of total youth-person-hours spent working on the farm/garden per year
- # of youth the farm/garden trained in job skills
- #,% of youth who participated in a youth program that graduate from high school
- #,% of youth who participated in a program that report high self esteem

#, % of youth who participated in a program reporting they have at least one good relationship with an adult other than their parent (mentor)

#,% of youth indicating positive attitude change and/or aspirations related to participating in youth-adult partnerships

#,% of youth who participated in a program reporting that they are/identify with being "part of their food community"

#,% of individuals reporting positive attitude and/or aspiration related to volunteering for civic activities, community service, and/or philanthropy

#,% of youth engaged in the political/governance process

Food access

#($),% of sales from food access programs

# of CSAs linked to the farm/garden

# of participants in the CSA linked to the farm/garden

#,% low income shares in CSA linked to the farm/garden

(pounds) of food produced by the farm/garden

of people participating in programs/activities that are growing food because they face food security issues

Age integrated spaces

of participants over 65

Economic

Urban agriculture can help stimulate local economies, provide jobs and job training, and offer opportunities to earn income through sales of produce and other agricultural products.

Local and regional economic stimulation

Unless otherwise noted, data can be collected through a tracking form.

$ total value of food produced

$ total value per square foot of produce

#, ($) total revenue generated from sales of food**

% of total revenue from sales of regional produce

% of CSA sales that is from regional produce

(hours) of volunteer-time contributed to the farm/garden

#, ($) from urban agriculture-based farmers market sales

% increase in urban agriculture-based farmers market sales from prior year

** Note: For some farms and gardens, particularly community farms, maximizing revenue is not a top priority. This indicator does NOT demonstrate that some farms or gardens are more efficient than others.

Job growth

of people employed by the farm/garden

#,% of new jobs created by the farm/garden

Job readiness

of people the farm/garden has trained in job skills

# of youth the farm/garden trained in job skills

#,% of farm/garden programs that employ youth

Affordable healthy food

#($),% of sales from food access programs

Ecological

Urban agriculture sites can contribute to stormwater capture, soil remediation and enrichment, and biodiversity. Additionally, urban agriculture can promote stewardship and community-led planning, as well as a shift toward local and regional food systems.

Awareness of food systems ecology[46]

Unless otherwise noted, data can be collected through a tracking form.

Indicators marked in color require special survey or study.

# of school students participating in food system ecology programs

Stewardship

of total participants in food system ecology programs

(pounds) of food produced by the farm/garden

(pounds) of food produced per square foot

community perceptions of the importance of urban agriculture as part of green/open space

Conservation

(square footage) of rooftop area collected for rainwater harvesting

#($) annual consumption of water use

#($) annual consumption of energy

#,% (square footage) of land/lot that could potentially grow food

#, % (square footage) of land/lot actually used to grow food

(pounds) of food waste processed (for compost)

Stormwater management

(square footage) of permeable surface in the farm/garden

Soil improvement

lead level in the farm's/garden's soil, per year

(pounds) of compost produced/processed

Reducing food waste

(pounds) of food waste collected

Habitat improvement/biodiversity/ecological connectivity

of activities related to increase biodiversity

# of beehives that are part of the farm/garden

# of chickens that are part of the farm/garden

% of vegetation planted with native/heirloom varieties through the farm/garden

# of crops grown

% of produce grown (as measured by % of total weight) without use of synthetic pesticides

% of produce grown (as measured by % of total weight) without use of synthetic fertilizers

46 Food system ecology applies ecological concepts and principles to the design, development, and management of sustainable agricultural systems.

37 D'Abundo, M., & Carden, A. (2008). Growing wellness: The possibility of collective wellness through community garden education programs. Community Development, 39, 4, 83-94. doi:10.1080/15573330809489660; McAleese, J. D., & Rankin, L. L. (2007). Garden-based nutrition education affects fruit and vegetable consumption in sixth-grade adolescents. Journal of the American Dietetic Association, 107, 662–665; Heim, S., Stang, J., & Ireland, M. (2009). A garden pilot project enhances fruit and vegetable consumption among children. Journal of the American Dietetic Association, 109, 1220–1226; Hermann, J. R., Parker, S. P., Brown, B. J., Siewe, Y. J., Denney, B. A., & Walker, S. J. (2006). After-school gardening improved children's reported vegetable intake and physical activity. Journal of Nutrition and Education Behavior, 38, 201–202; Lawson, L. (2007). Cultural geographies in practice: The South Central Farm: Dilemmas in practicing the public. Cultural Geographies, 2007, 611–616; Lineberger, S. E., & Zajicek, J. M. (2000). School gardens: Can a hands-on teaching tool affect students' attitudes and behaviors regarding fruit and vegetables? HortTechnology, 10, 593–597; McCormack, L. A., Laska, M. N., Larson, N. I., & Story, M. (2010). Review of the nutritional implication of farmers markets and community gardens: A call for evaluation and research efforts. Journal of the American Dietetic Association, 110, 399–408; Morris, J. L., & Zidenberg-Cherr, S. (2002). Garden-enhanced nutrition curriculum improves fourth-grade school children's knowledge of nutrition and preferences for some vegetables. Journal of the American Dietetic Association, 102, 91–9; Robinson-O'Brien, R., Story, M., & Heim, S. (2009). Impact of garden-based youth nutrition intervention programs: A review. Journal of American Dietetic Association, 109, 273–280.

38 Alaimo, K., Packnett, E., Miles, R. A., & Kruger, D. J. (2008). Fruit and vegetable intake among urban community gardeners. Journal of Nutrition Education and Behavior, 40, 94–101.

39 Austin, E. N., Johnston, Y. A. M., & Morgan, L. L. (2006). Community gardening in a senior center: A therapeutic intervention to improve the health of older adults. Therapeutic Recreation Journal, 40, 48–56; Armstrong, D. (2000a). A community diabetes education and gardening project to improve diabetes care in a northwest American Indian tribe. Diabetes Educator, 26, 113–120; Hannah, A. K., & Oh, P. (2000). Rethinking urban poverty: A look at community gardens. Bulletin of Science, Technology and Society, 20, 207–216; Twiss, J., Dickinson, J., Duma, S, Kleinman, T., Plausen, H., & Rilveria, L. (2003). Community gardens: lessons learned from California healthy cities and communities, American Journal of Public Health, 93, 1435–1438.

40 French SA and Wechsler H. 2004. "School-based research and initiatives: fruit and vegetable environment, policy, and pricing workshop." Prev Med. 2004 Sep;39 Suppl 2:S101-7.

41 Teig, E., Amulya, J., Bardwell, L., Buchenau, M., Marshall, J. A., & Litt, J. S. (2009). Collective efficacy in Denver, Colorado: Strengthening neighborhoods and health through community gardens. Health & Place, 15, 1115–1122; D'Abundo, M. L., & Carden, A. M. (2008). "Growing wellness": The possibility of promoting collective wellness through community garden education programs. Journal of the Community Development Society, 39, 83–94; Hannah, A. K., & Oh, P. (2000). Rethinking urban poverty: A look at community gardens. Bulletin of Science, Technology and Society, 20, 207–216; Lawson, L. (2007). Cultural geographies in practice: The South Central Farm: Dilemmas in practicing the public. Cultural Geographies, 2007, 611–616; Macias, T. (2008). Working toward a just, equitable, and local food system: The social impact of community-based agriculture. Social Science Quarterly, 89, 1086–1101; Roubanis, J. L. & Landis, W. (2007). Community gardening project: Meredith College students explore sustainability, organics. Journal of Family and Consumer.

42 Ferris, J., Norman, C., & Sempik, J. (2001). People, land and sustainability: Community gardens and the social dimension of sustainable development. Social Policy and Administration, 35, 559–568; Krasny, M. E. & Tidball, K. G. (2009). Community gardens as contexts for science, stewardship, and civic action learning. Cities and the Environments, 2, 1–18; Pudup, M. B. (2008). It takes a garden: cultivating citizen-subjects in organized garden projects. Geoforum, 39, 1228–1240; Rahm, J. (2002). Emergent learning opportunities in an inner-city youth gardening program. Journal of Research in Science Teaching, 39, 164–184.

43 Allen, J. O., Alaimo, K., Elam, D., & Perry, E. (2008). Growing vegetables and values: Benefits of neighborhood-based community gardens for youth development and nutrition. Journal of Hunger & Environmental Nutrition, 3, 418–439; Blair, D. (2009). The child in the garden: An evaluative review of the benefits of school gardening. Journal of Environmental Education, 40, 15–38; Doyle, R., & Krasny, M. (2003). Participatory rural appraisal as an approach to envi- ronmental education in urban community gardens. Environmental Education Research, 9, 91–115; Fusco, D. (2001). Creating relevant science through urban planning and gardening. Journal of Research in Science Teaching, 38, 860–877; Graham, H., & Zidenberg-Cherr, S. (2005). California teachers perceive school gardens as an effective nutrition tool to promote healthful eating habits. Journal of the American Dietetic Association, 105, 1797–1800; Krasny, M. E. & Tidball, K. G. (2009). Community gardens as contexts for science, stewardship, and civic action learning. Cities and the Environments, 2, 1–18; Lautenschlager, L., & Smith, C. (2007a). Understanding gardens and dietary habits among youth garden program participants using the Theory of Planned Behavior. Appetite, 49, 122–130; Ozer, E. J. (2007). The effects of school gardens on students and schools: Conceptualization and consideration for maximizing healthy development. Health Education and Behavior, 34, 846–863; Rahm, J. (2002). Emergent learning opportunities in an inner-city youth gardening program. Journal of Research in Science Teaching, 39, 164–184; Robinson-O'Brien, R., Story, M., & Heim, S. (2009). Impact of garden-based youth nutrition intervention programs: A review. Journal of American Dietetic Association, 109, 273–280.

Phoenix Community Garden, Brooklyn

III. POLICY

INTRODUCTION

"Right now urban agriculture's on everyone's lips and it sounds good. But yet, are people in power then making policy? Are they thinking of it as, 'Oh, it's a new wave and it will go away'? Or do they really believe in what they're saying: as the city goes into the next decade or so and even further, that community gardens and urban farms must be part of the landscape when it comes to urban planning?"

– Community Gardener

Policy decisions affect every aspect of urban agriculture, from the availability of soil and growing space to the cost of running a farmers market and the extent to which children learn about food and agriculture in their schools. Fortunately, there is no shortage of good policy ideas to support and expand urban agriculture in New York City. Recent legislation and agency actions have sought to make it easier to build rooftop greenhouses and have funded urban farms and gardens. Several policy documents, including the Mayor's citywide sustainability plan, PlaNYC, and the City Council Speaker's comprehensive food policy platform, FoodWorks, have asserted the importance of urban agriculture to community development, food access and open space, recommending dozens of initiatives to support the city's farmers and gardeners.

Despite this flurry of policy activity, numerous stakeholders interviewed for this project—including government officials themselves—lacked confidence that City government is committed to perpetuating urban agriculture as a wide-scale use of public space, particularly over the long term. These doubts stem from several factors. First, there is currently no Mayoral directive or Local Law to promote urban agriculture. Lacking this legally binding mandate, government agencies have limited resources and authority to address urban agriculture, incorporate it into potentially complementary plans and operations, or coordinate activity with other agencies. Farmers and gardeners interviewed described in very positive terms the help they receive from municipal programs such as GreenThumb, but noted that their ability to obtain services or information from government agencies has been uneven, and that many government regulations and processes are unclear. For instance, government officials have no official policy for evaluating proposals for new farms and gardens on City-owned land or for making compost produced in New York City available to farmers and gardeners.

This chapter recommends government policies and practices that would integrate urban agriculture more fully into New York City's physical landscape and agency procedures, making farms and gardens as much a part of the city's sustainability fabric as waste recycling, transportation infrastructure, water and sewer service, parks and open space, and community development programs. The proposed recommendations would make the city's urban agriculture system—from the allocation of growing space and the delivery of services to the coordination of numerous stakeholders' actions—more efficient, transparent, and participatory, and as a result, better able to achieve key municipal goals. These recommendations include the following:

- Develop an urban agriculture plan that establishes goals, objectives, a citywide land use scheme for garden and farm development, and adequate agency budgets to support existing and future urban agriculture activity.

- Integrate urban agriculture into existing plans, programs, and policy-making processes in city government, including the Department of Environmental Protection's Green Infrastructure Program and the Department of Sanitation's plans for compost production, and identify opportunities for existing initiatives to achieve multiple goals while supporting farmers and gardeners.

- Foster innovative opportunities to build urban agriculture into the cityscape, from new housing complexes and existing rooftops, to sidewalks and stalled development sites.

- Address disparities in access to funding, information, and other resources by creating more transparent and participatory processes—such as a citywide Urban Agriculture Task Force—to enable gardeners and farmers to influence policy and decision-making.

- Address race- and class-based inequities by supporting capacity building among underserved groups and within city agencies.

- Make existing administrative processes more responsive to urban agriculture constituents, making it easier for enterprising farmers and gardeners to thrive.

The recommendations consider policy at different scales, from citywide plans and agency-specific regulations to decision-making that affects individual sites. They address not only food production, but also services and infrastructure—from compost production to retail distribution channels—that support urban agriculture. They propose changes to specific laws and regulations, as well as to the processes and organizational structures that determine how agency priorities are established and decisions are made. Finally, although the recommendations focus on the roles of government agencies and elected officials, these stakeholders should partner whenever possible with philanthropies, support organizations, private businesses, universities, farmers and gardeners, and the many other advocates for urban agriculture in New York City.

Among the many actions recommended in this chapter, two in particular would better equip city government to address urban agriculture citywide. The first is to clearly define and fully fund leadership positions and staff to develop urban agriculture policy and provide services to farmers and gardeners. Bolstering the budget and staff of GreenThumb, a division within the Department of Parks & Recreation that provides technical and material assistance to hundreds of farms and gardens in all five boroughs, would help ensure that these sites receive critical support, and would facilitate the creation of new urban agriculture projects. In addition, the Food Policy Coordinator, who currently convenes multiple agencies to address the city's food system, is well-positioned to work with the many agencies currently involved with urban agriculture to develop policies to address the availability of City-owned land, soil, and other resources needed by urban farmers and gardeners, and to develop a citywide urban agriculture plan.

An urban agriculture plan would identify locations for farms, gardens, and urban agricultural infrastructure, such as commercial kitchens, farmers markets, and composting sites. It would engage the public, allowing farmers, gardeners and other urban agriculture stakeholders to lend their expertise in shaping policy, and enabling them to document the current and potential impacts of urban agriculture. An adopted plan also would have the legal status that existing policy reports and even PlaNYC lack. With a plan in place, the City would have an opportunity to capture the full economic, ecological, health, and social benefits of a larger and more robust network of gardens and farms, and the myriad entrepreneurial ventures that New Yorkers will create.

EVOLUTION OF URBAN AGRICULTURE POLICY

The goal of providing space for a crowded, expanding population to grow food in New York City dates back to the early 20th century. Dozens of school gardens and farm plots in public parks flourished through the 1930s. In 1937, Parks officials reported that over 1.2 million pounds of vegetables were harvested on parkland. Through initiatives such as the U.S. School Garden Army, Women's Land Army of America, and Victory Gardens, government agencies promoted agriculture in cities to alleviate the effects of urban poverty and contribute to national food security, particularly during times of war.[47] Less emphasis was placed on maintaining gardens in public space during the 1950s, as the national food supply stabilized, suburban development surged, and backyard landscaping became a focus of new homeowners and the horticultural industry.[48]

During the late 1960s and early 1970s, a time of fiscal crisis and social unrest in New York and other major U.S. cities, urban agriculture activity increased nationwide. In New York, where housing abandonment and reduced municipal services devastated many neighborhoods, residents began transforming vacant lots into vibrant community gardens, including spaces to grow food.[49] Community members, rather than government agencies, typically led these projects, many of which focused explicitly on social justice and "self-help."[50] New support organizations were formed: for example, New York City's Green Guerillas was founded in 1973 to help residents organize the cleaning and planting of vacant lots for food production, neighborhood revitalization, and grassroots development in terms of both the physical landscape and community empowerment. New York City's government soon followed the lead of community efforts. In 1978, the Koch administration created "Operation Green Thumb" (now called GreenThumb) to provide technical support to community gardeners and to assist in the management of city-owned garden sites. The Garden and Greening Program of the New York City Housing Authority (NYCHA) and an urban agriculture technical assistance program within Cornell Cooperative Extension also provided municipal support for urban agriculture.

The urban agriculture community was galvanized in 1999, when the Giuliani administration attempted to auction off the land occupied by 115 community gardens to housing developers, setting off legal challenges and protests, and drawing the intervention of the state Attorney General. For many farmers and gardeners, Giuliani's efforts highlighted the vulnerability of farm and garden space in a city that was rebounding economically, as well as the importance of engaging in the process of shaping public policy.

Over the last decade, urban agriculture has expanded across the country, as city residents—particularly in low-income neighborhoods—increasingly recognize the many functions that community gardens and farms serve, including providing access to affordable fresh, healthy, locally produced food.[51] Planners and policy makers have responded to these trends by attempting to weave agriculture back into the urban landscape, revising their local zoning codes, ordinances, and development policies to accommodate, regulate, and support urban agriculture activities.[52]

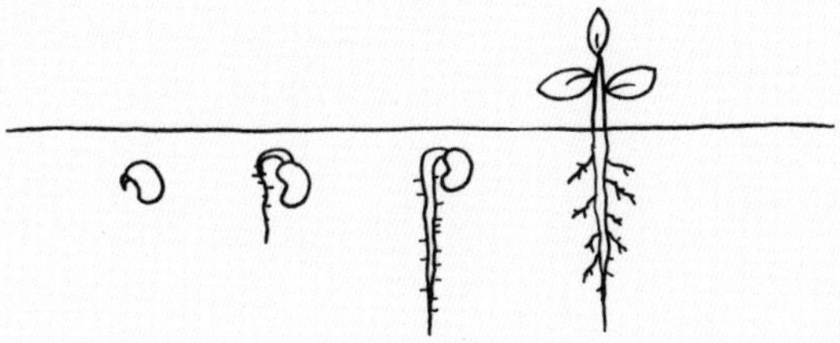

47 Lawson, Laura J. 2005. op cit.; Hayden-Smith, Rose. 2006. Soldiers of the soil: A historical review of the United States

48 Hynes, P.H. 1996. op. cit., p. xiii-xiv; Lawson, Laura J. 2005 op cit, p 205-207.

49 Lawson, Laura J. 2005. op cit., p. 219.

50 Lawson, Laura J. 2005. op cit., p 206-208; Hynes, P.H., and G. Howe. 2002. Urban horticulture in the contemporary United States.

51 Martinez, S., et al. (2010). Local Food Systems: Concepts, Impacts, and Issues. Economic Research Service 97. Washington, DC: US Department of Agriculture; Marketing of local foods grossed $4.8 billion in 2008. See Low, Sarah A., and Stephen Vogel. Direct and Intermediated Marketing of Local Foods in the United States, ERR-128, U.S. Department of Agriculture, Economic Research Service, November 2011.

52 Hodgson, K., et al., op cit.; Viljoen, A., Bohn, K., & Howe, J. (2005). Continuous Productive Urban Landscapes: Designing Urban Agriculture for Sustainable Cities. Oxford: Architectural Press; Nordahl, D. (2009). Public Produce. Washington, DC: Island Press.

Recent New York City Initiatives

Five Borough Farm builds on the momentum to adopt policies supporting urban agriculture citywide, particularly on the analysis and proposals outlined in four recent documents: the update to PlaNYC, the City's sustainability plan; FoodWorks, a report issued by the Speaker of the City Council; FoodNYC, a white paper published by the Manhattan Borough President; and "The Potential of Urban Agriculture in New York City," a report published by Columbia University's Urban Design Lab.

PlaNYC

On April 22, 2007, the Bloomberg administration launched a citywide sustainability plan, PlaNYC.[53] That document did not mention urban food production. However, an update to PlaNYC released in April 2011 states that sustainable food systems are critical to the city's well-being and included food as an issue that requires actions from multiple agencies. In particular, the updated PlaNYC states:

> We are committed to promoting community gardens and other forms of urban agriculture. We recognize the important role they serve in building communities, supporting local cultural heritage, and bringing individuals together around the vital issue of access to healthy food.[54]

PlaNYC also acknowledges that urban agriculture provides open space in communities with few or no formal parks. For example, in the Parks and Open Space section, PlaNYC states:

> We will target high-impact projects in the neighborhoods with the greatest open space needs. These projects will include community gardens and urban agriculture opportunities, which enrich many of the city's neighborhoods least served by parks.[55]

Finally, PlaNYC outlines several specific objectives to increase urban agriculture:

- The New York City Housing Authority will expand its community gardening program to include the creation of "at least one urban farm" as well as 129 new community gardens on Housing Authority land.

53 http://www.nyc.gov/portal/site/nycgov/menuitem.c0935b9a57bb4ef3daf2f1c701c789a0/index.jsp?pageID=mayor_press_release&catID=1194&doc_name=http%3A%2F%2Fwww.nyc.gov%2Fhtml%2Fom%2Fhtml%2F2007a%2Fpr119-07.html&cc=unused1978&rc=1194&ndi=1

54 Ibid, pg. 37.

55 PlaNYC. (2011) pg. 35.

- The Department of Parks & Recreation will launch a study to "identify potential urban agriculture or community garden sites on City-owned properties unsuitable for other development." In addition, the Department will increase the number of community volunteers registered with GreenThumb by 25 percent, expand support for community gardens into new underserved neighborhoods, and establish five new farmers markets at community garden sites.

- The Mayor's Fund and the Department of Education will register 25 new school gardens per year, retaining 75 percent of new registered school gardens year to year.

- The Departments of City Planning, Buildings, and Parks & Recreation will review laws and regulations and take steps to reduce existing regulatory barriers to urban agriculture.

- The Department of Sanitation will reinstate leaf and yard waste composting, expand composting of park leaf and grass clippings, and evaluate the feasibility of a curbside organic waste composting program.

PlaNYC is a strategy document that reflects the vision and objectives of the administration and the Office of Long-Term Planning and Sustainability, which prepares the document. However, at this time there is no requirement in the City Charter or Administrative Code for agencies to ensure that their policies, plans, or actions conform to the goals and targets outlined in PlaNYC, nor does PlaNYC commit the city to a particular course of action.

Other Policy Documents

Two policy documents prepared by elected officials with input from key stakeholders, experts, and the general public indicate that urban agriculture is important to the city's future and offer recommendations to support and grow the city's gardens and farms: FoodWorks, a comprehensive, citywide food policy platform issued by City Council Speaker Christine C. Quinn in November 2010, and "FoodNYC: A Blueprint for a Sustainable Food System," a white paper issued by Manhattan Borough President Scott Stringer in February 2010. In addition, a recent report from Columbia University's Urban Design Lab highlights opportunities to expand urban agriculture citywide, and also includes a set of recommendations.

Collectively these documents have helped build public support for policies to strengthen urban agriculture citywide, and have advanced specific proposals that stakeholders can explore or promote. Most notably, FoodWorks has guided the City Council to enact new local laws that mandate specific agency actions to address urban agriculture.[56]

---

56 Ackerman, K. (2011) The Potential for Urban Agriculture in New York City: Growing Capacity, Food Security, and Green Infrastructure. NY: Urban Design Lab at the Earth Institute.

Implementation of FoodWorks Strategies

Shortly after releasing FoodWorks, the City Council passed several new local laws and adopted a resolution to implement a number of urban agriculture-related policies. They include:

- Local Law 48 of 2011, which requires the Department of Citywide Administrative Services (DCAS) to maintain an online database of all property owned and leased by the City, including detailed data about the sites as well as whether land is potentially suitable for urban agriculture. The database has been made public, but information on which sites are suitable for urban agriculture has not yet been included.

- Local Law 49 of 2011, which adds greenhouses to the list of rooftop structures that can be excluded from building height limitations, making it easier to install the structures atop existing buildings.

- Local Law 52 of 2011, which requires the Mayor's Office of Long-Term Planning and Sustainability to gather and report on key data about New York City's food system, including sources of food purchased by the city (including, potentially, community gardens), and how food is distributed and consumed. An annual food system metrics report is due September 1, 2012, and annually thereafter.

- Resolution 527, which calls on the State Legislature to extend the Green Roof Tax Abatement to live food-producing plants as well as low-maintenance sedums, thus encouraging more property owners to install rooftop farms.

- The zoning text amendment approved by the City Council that allows greenhouses to be exempt from floor area and height limits on commercial buildings.

RECOMMENDATIONS

1. Formalize City Government's Support for Urban Agriculture

City government has responded to the tremendous surge in urban agricultural activity in many ways. Agencies have adopted ad hoc agreements with individual farms and gardens, for instance, and enterprising staff have stretched their resources and official mandates to support urban agriculture. Policy makers have developed many recommendations that would facilitate the expansion of urban agriculture in New York. Government officials now have an opportunity to coordinate these efforts on a citywide scale, to codify in law the goals and strategies that have been outlined in documents such as PlaNYC and Food-Works, and to demonstrate the city's long-term commitment to urban agriculture.

1.1 Establish a clear urban agriculture policy

A first step to integrate urban agriculture formally in the city's governance is to establish an urban agriculture policy in the Administrative Code or City Charter. Changing the Administrative Code or Charter requires enactment of a local law, enabling a wide range of stakeholders to provide input, and requiring affirmation by a majority of the City Council and the Mayor.[57]

Such a policy would indicate that urban agriculture is an important element in the city's sustainable food system that contributes to social, health, economic, and ecological benefits. As in PlaNYC, a policy statement would outline goals, such as supporting urban agriculture as an open space amenity in underserved neighborhoods, making school gardens a year-round community resource, or promoting economic development opportunities for low-income residents at urban farms and gardens. It would also set measurable objectives, such as creating a certain number of new farms and gardens, or creating a certain number of new urban agriculture jobs. Finally, it would specify the roles and responsibilities of City agencies in achieving these goals and objectives.

A policy commitment would have several effects. It would:

- Empower agency commissioners to initiate new programs, institute supportive agency practices, push for expanded budgets to run programs, and introduce regulations to advance urban agriculture.

- Legitimize agency staff support for urban agriculture, especially in those agencies not directly responsible for gardens and farms.

57 The city has established public policy goals, such as its commitment to waste recycling, in the Administrative Code. See NYC Administrative Code § 16-302: Declaration of policy. It is hereby declared to be the public policy of the city to reduce environmental pollution and dangers to health, to decrease the demand for scarce landfill space, to minimize the size and cost of the proposed resource recovery program, and to encourage the conservation of valuable natural resources and energy. It is the policy of the city to promote the recovery of materials from the New York City solid waste stream for the purpose of recycling such materials and returning them to the economy.

LESSONS FROM SEATTLE

In Seattle, a City Council resolution and clear commitments from the Mayor in support of urban agriculture have fostered greater collaboration across departments and enabled urban agriculture to be treated as a priority within specific agencies.

Seattle's active network of community gardeners, environmentalists, and academics has worked to develop and support the implementation of urban agriculture policy. In 2008, their support helped lead the Seattle City Council to adopt Resolution Number 31019, which established food system goals, including proposed policies to support the expansion of urban agriculture.[12] The Council's food system resolution identified the roles of multiple city agencies in supporting a sustainable food system: for instance, it directs the Department of Neighborhoods to identify community kitchens and other facilities to support community gardens, and the Transportation Department to consider pedestrian, bicycle, and transit connections between residential neighborhoods and community gardens, food banks, food markets, and farmers markets in evaluating new transportation projects. The resolution was accompanied by a ballot initiative in November 2008, in which 59 percent of Seattle voters approved a Parks and Green Spaces Levy that provided $2 million in funding for the development of new community gardens.[13]

Integrating urban agriculture in the missions of different Seattle agencies has led to innovative policy development. For example, the Seattle Public Utilities department created a backyard composting project, has extensive public information on resource-conserving gardening in the city, and also supports food recovery from businesses for both waste reduction and food access. In all major land use decisions, the Department of Planning and Development is considering the potential benefits and impacts of its actions on the city's food system.

In February 2010, Mayor Mike McGinn and the City Council declared 2010 "the year of urban agriculture," launching a campaign to promote urban food production and access to locally grown food. **The Mayor has directed multiple agencies to work together to support urban agriculture, establishing an interagency food systems team comprising eight city agencies** (including Public Works, Seattle City Power and Light, Parks and Recreation, and the Office of Economic Development). Parks and Recreation is working with the Department of Neighborhoods to integrate community centers into the city's community garden program and provide community kitchen facilities to neighborhood groups. The City's sustainable building initiative, a joint Seattle Public Utilities and Department of Planning and Development project, has been exploring how its green roof program can promote rooftop urban agriculture. In August 2010, the Seattle City Council adopted land use code changes to recognize and allow five different types of urban agriculture land uses.[14]

The key political factor behind the development of urban agriculture policy in Seattle was the support of the McGinn administration, together with interested council members, and the work of advocates and gardeners to build support among Seattle voters, bolstered by city policies.

---

58 http://clerk.seattle.gov/~scripts/nph-brs.exe?s1=&s2=&s3=31019&s4=&Sect4=AND&l=20&Sect2=THESON&Sect3=PLURON&Sect5=RESN1&Sect6=HITOFF&d=RES3&p=1&u=%2F~public%2Fresn1.htm&r=1&f=G\

59 http://www.cityofseattle.net/parks/levy/

60 http://tinyurl.com/26z5fzq, accessed on December 10, 2010

- Signal to entrepreneurs, the philanthropic community, educational institutions, and businesses that urban agriculture has a future in New York, thus encouraging private investment in new and expanding food production and social enterprise ventures.

- Help advocates within and outside of government maintain City budgets for urban agriculture, even if priorities change from one Mayoral administration to the next.

But a policy commitment is only the first step toward institutionalizing urban agriculture. It should be followed by the development of an urban agriculture plan.

1.2 Develop an urban agriculture plan

A citywide urban agriculture plan would build on the goals, objectives, and agency roles established in an urban agriculture policy statement, and would include many dimensions not addressed in a policy statement.

First, an urban agriculture plan would gather and analyze data on land uses, economic, social, and environmental conditions, and expected trends, to specify where new urban farms and gardens should be located. Second, it would be developed with public participation, engaging a wide range of stakeholders to shape the plan. Third, unlike strategy documents or policy papers, an urban agriculture plan would be approved through procedures in the City Charter.[61] Fourth, adopted plans are meant to be considered alongside changes to zoning and other agency plans[62] and programs, such as community-based plans and citywide plans for comprehensive solid waste management,[63] composting,[64] and sustainable stormwater management.[65] Finally, an urban agriculture plan would transcend individual administrations, and, with regular updates, would evolve as the city's food system, economy, priorities, and technologies change.

Specifically, an urban agriculture plan should:

- Determine the extent to which urban agriculture should be expanded, based on an assessment of the social, economic, and ecological benefits associated with farming and gardening, and factors such as the potential capacity for food production, the costs of expanding production, opportunity costs for using land for agriculture, and the potential for urban agriculture to meet the needs of different communities.

---

61 For instance, Section 197-a of the New York City Charter provides for approval of long-term plans, including citywide plans.

62 See New York City Department of City Planning. Community Based Planning. The 197-a Plan. Accessed at http://www.nyc.gov/html/dcp/html/community_planning/197a.shtml on January 19, 2012.

63 http://www.nyc.gov/html/dsny/html/swmp/swmp-4oct.shtml

64 Laws of New York City § 16-316.2 requires a food waste composting plan. The Sanitation Department, in conjunction with the mayor's office of long-term planning and sustainability, is required to issue a report by 7/1/2012 recommending methods to expand the diversion of compostable waste from the city's waste stream.

65 See http://www.nyc.gov/html/dep/html/stormwater/nyc_green_infrastructure_plan.shtml

- Identify, with extensive citizen input, the neighborhoods that would benefit the most from urban agriculture–related activities and benefits, including youth leadership programs, garden-based educational opportunities, and access to healthy food.

- Assess the land and rooftop space needs for a variety of urban agriculture scenarios over the next decade, including identifying specific parcels of land and rooftops that should be used for urban agriculture in the future.

- Determine annual operating budgets, infrastructure, and long-term capital budget needs to adequately support the city's urban agriculture program.

- Create processes that ensure coordination across City agencies (and between city and state/federal agencies) to support urban agriculture.

- Require City agencies to consider urban agriculture in their development of plans. For example:

    - Updates to the Comprehensive Waterfront Plan could identify waterfront sites that are appropriate for food production and distribution.

    - School Wellness Plans could give parents a process to include gardening as a part of their school's wellness programs.

    - The Citywide Solid Waste Management Plan could include strategies to meet the compost needs of the city's gardens and farms.

Making Commitments to Urban Agriculture in City Plans: Seattle and Vancouver

In Seattle, the broad goals of the city's P-Patch community gardening program are tied to specific standards for the appropriate amount of space devoted to urban agriculture per capita. **Seattle's 2005 comprehensive plan calls for at least one community garden for every 2,500 households in an urban village or neighborhood** (Seattle Comprehensive Plan, Urban Village Appendix B).[65] Vancouver's Greenest City 2020 Action Plan not only sets forth the broad aspiration to make the Canadian city a global leader in urban food systems, but also establishes a specific target of increasing city and neighborhood food assets (including garden space) by at least 50 percent from 2010 levels. Among the highest-priority actions for the next three years are creating five to six community gardens per year, enabling three new urban farms, adding public fruit trees, and developing other food processing and distribution infrastructure. And the plan designates a lead agency—Social Policy—responsible for implementation, with support from two other key agencies, Engineering and Parks.[66]

65 Available at http://www.seattle.gov/dpd/cms/groups/pan/@pan/@plan/@proj/documents/Web_Informational/cos_004510.pdf

66 City of Vancouver. Administrative Report RR-1. July 5, 2011. Accessed at http://vancouver.ca/ctyclerk/cclerk/20110712/documents/rr1.pdf on January 27, 2012

Minneapolis' Urban Argriculture Plan

In February 2011, Minneapolis adopted an Urban Agriculture Policy Plan, which makes a number of recommendations to support urban agriculture. The plan calls on the city to "prioritize local food production and distribution" when deciding on the use of city-owned and private property, including new development projects "that could potentially affect existing local food resources." It also requires the city to integrate farmers markets into development plans, identify policies to encourage green roofs for food production, and create incentives for developers to include space for food production, distribution, and composting in new projects. Such policies include allowing urban agriculture to count toward green space set-aside requirements and green building requirements.[67]

The Minneapolis plan also calls for the creation of an "overarching policy framework" to support urban agriculture, including an inventory of land for agriculture and food distribution, policies to support ownership or long-term tenure for growers and farmers markets, policies to reduce liability and property taxes for urban farms and distribution facilities, and policies that make vacant and foreclosed properties more accessible for food growing and distribution. Following the adoption of the plan, the Minneapolis City Council approved a zoning code text amendment on January 23, 2012, to allow urban agriculture uses within the city.[68]

    - PlaNYC could address the role of urban agriculture on issues such as neighborhood sustainability, management of waste and stormwater, and public health.

- Track and evaluate the effectiveness of City agency activities to support urban agriculture, and ensure that such information is used to make continuous improvements. This could include developing agency urban agriculture indicators to be included in the Mayor's Management Report so that agencies, policy makers, and citizens can gauge agency performance with respect to urban agriculture;

- Improve governance structures, practices, and programs to broaden participation in urban agriculture policy-making and planning, and provide more equitable access to material and financial resources.

The Food Policy Coordinator, working with the Department of City Planning and other City agencies, could lead the production of a plan including the compilation and analysis of existing data, and public participation processes. While some City officials have expressed concern about the potential cost and time of preparing such a plan, which is not currently funded in the City's budget, a number of tasks required to complete a plan are already mandated, under way, or completed. For instance, as noted above, Local Law 48 of 2011 requires the DCAS to identify city-owned parcels that are potentially suitable for urban agriculture, and data on more than 1,000 GreenThumb and public housing gardens are already collected by the Department of Parks & Recreation and the New York City Housing Authority.

67 Minneapolis was able to prepare an urban agriculture plan in-house for $150,000, including a $50,000 consultant contract and approximately $100,000 for city agency staff. (Personal e-mail exchange with Amanda Arnold, AICP, Principal City Planner, City of Minneapolis, on January 18, 2012.)

68 http://www.minneapolismn.gov/meetings/planning/WCMS1P-085268

The following are examples of key milestones in creating an urban agriculture plan:

1.2.1 – Create an agriculture land use map

An urban agriculture land use map, prepared with community input, would document the size, location, citywide distribution, and status of sites that might be appropriate for urban agriculture. This document would help determine how many parcels overall could feasibly be dedicated to urban agriculture, how agricultural land should be distributed throughout the five boroughs, and how long land should be dedicated to urban agriculture. It would also help evaluate where urban agriculture–related infrastructure such as food processing facilities (e.g., to wash, cut, and bag produce) and food distribution hubs could be strategically located to take advantage of the food produced on urban farms and gardens.

1.2.2 – Document all existing urban agriculture sites

There is no single list within City government that records the more than 700 farms and gardens growing food citywide. City agencies increasingly want to obtain information on urban agriculture, but slim agency budgets, a lack of staff assigned to address urban agriculture, and limited capacity to manage large amounts of data housed within multiple agencies overseeing farms and gardens make it difficult to document urban agricultural activity citywide.

Local Law 52 of 2011 requires the Mayor's Office of Long-Term Planning and Sustainability to publish a report on the city's food system by September 2012 that includes the location (sorted by community board and size in square feet) of each community garden on city-owned property that is registered with and licensed by the Department of Parks & Recreation, and whether each garden engages in food production. While this is a step in the right direction, **the Mayor's Office of Long-Term Planning and Sustainability should compile and publish an online registry of all urban agriculture sites in New York City, including community farms, school gardens, public housing gardens, institutional farms, and commercial farms.**

In addition to the information required by Local Law 52, this database should also include the following:

- Square feet being gardened or farmed (not merely the total square feet of the garden or farm lot)
- The type and quantity of food being produced
- Related activities that take place on the site
- The expected or desired duration of farming and gardening on each site

The Mayor's Office of Long-Term Planning and Sustainability could outline steps and provide resources to assist farmers and gardeners in collecting this data, which would help City officials anticipate future needs for urban agriculture infrastructure and determine where farming activities are likely to increase or decrease across the city. As part of the land use planning process for urban agriculture, City officials would also be in the position to determine appropriate lease or license arrangements for farms and gardens that should have extended protection from development, and sites that should be farmed or gardened on an interim basis.

In a city as large as New York, with new urban agriculture projects coming on-line regularly, even a well-funded and -staffed effort to compile growing sites will quickly become outdated without frequent updates. **City officials, working with funders and support organizations, should develop an interactive database that could be updated by urban farmers and gardeners, and the organizations that support them.** GreenThumb, which already conducts regular visits to gardens citywide, could spot-check sites added by the public. Creating this interactive database would maximize limited city resources and tap the on-the-ground knowledge of community members. It also would signal to the urban agriculture community that they are partners in developing city policy.

1.2.3 – Document available city-owned property

NYC OpenData, a City government online database, provides information on the approximately 18,500 city-owned parcels that are currently used by City agencies for public buildings (such as schools, police stations, libraries, and warehouses); open spaces (such as piers, parks, and natural areas); highway maintenance yards and parking lots; or are unused and vacant. In the Local Law 48 of 2011 Report, users searching for vacant land can find the block and lot information, dimensions, inclusion in government clean up programs, existing structures, and other details of sites, including which agencies have jurisdiction over each site and contact information for further information.[69] DCAS has yet to meet the requirement in the law to assess any given site's suitability for urban agriculture. In fact, a provision in the law specifies that data must be provided to "the extent such information is available" to DCAS, which may relieve the agency of the obligation to gather new information. **The Council and advocates should work to ensure that this information is gathered by making available the funding and staff resources to complete an urban agriculture assessment and by tightening the requirements in the coming years.**

Identifying and Activating Vacant Land: 596 Acres

The nonprofit 596 Acres has created a website featuring a map of Brooklyn's vacant parcels and contact information for the City agency responsible for each parcel. The website also serves as a message board to help residents interested in reclaiming these lots for community use find each other. 596 Acres posts signs at the vacant lots to draw attention to, and educate passersby about, the possibility of turning those sites into gardens. The project has already resulted in the creation of new GreenThumb gardens and has enabled community members to organize to create others. While 596 Acres is no substitute for DCAS meeting the requirements of Local Law 48 of 2011, the project illustrates that gathering and conveying this information to the public can be done quickly, at a relatively low cost.

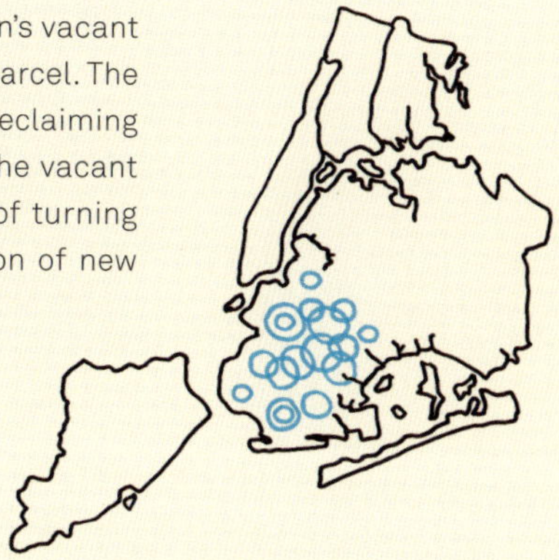

69 See the LL48 database: http://nycopendata.socrata.com/Government/Local-Law-48-Of-2011-Report/2b6x-2bw6

1.2.4 - Identify available private property

The NYC OpenData database covers City-owned property, but the City's agriculture land use map should also encompass only privately-owned open space and potential growing spaces atop flat-roofed buildings throughout the city.[70] A comprehensive citywide database that includes these private properties would help owners of vacant land or rooftop space connect with individuals and groups that are interested in using those spaces to grow food—either on a rental or sharecropping basis, or for free, depending on the owner's needs. Such a database could also be used to help agencies such as the Department of Environmental Protection identify participants for its Green Infrastructure Program, and enable entrepreneurs to identify parcels of land for commercial farming ventures.

1.2.5 – Develop criteria to evaluate the suitability of vacant land for urban agriculture

These criteria could include the following:

- Growing conditions, including exposure to sun, wind, etc.
- Quality of soils, including both soil fertility and toxicity
- Access to water and electricity
- Ease of accessing the site
- Compatibility with neighboring land uses and community goals for the site
- Need for open space in the community
- Nearby health or safety hazards
- Access to healthy food in the surrounding neighborhood
- Title or other legal obstacles to using the site
- Costs to convert the site to an agricultural use
- Value of alternative uses and development potential
    - Is the site intended to be developed?
      If so, how many years before construction will begin?
    - If not, is there market demand for development in the near future?
    - Is the parcel large enough to make it attractive to developers?[71]

70 Columbia University's Urban Design Lab has already conducted a preliminary assessment of private-owned spaces that potentially could be used for urban agriculture.

71 Developers may require lots to be a minimum size before they can build new apartments or commercial establishments, but urban farmers and gardeners can operate on very small sites—container gardens can fit into sites with as little as a few hundred square feet.

1.2.6 – Evaluate the availability and suitability of city-owned sites for agriculture

Once criteria to evaluate the suitability of city-owned land for urban agriculture have been developed, DCAS should review and apply these criteria to existing sites. DCAS could start with a manageable subset of its database of properties. Assessing even a few dozen sites would be a valuable way to estimate the time and costs of completing the evaluation of all vacant parcels. DCAS also could partner with community-based non-profit organizations that have identified suitable city-owned land and local residents interested in starting urban farms and gardens on those sites.

San Francisco Urban Agriculture Land Use Criteria

In San Francisco, a Mayoral Directive required all city agencies with jurisdiction over property to audit their land to identify parcels suitable for or actively used for food-producing gardens.[72] To do so, a group of agency representatives from the Mayor's Director of Greening, City Planning, Department of Public Health, and Department of the Environment developed the following evaluative criteria:

- Vacant or underused sites no less than 500 square feet with no portion less than 10 feet wide
- Availability for at least the next three years
- 30% of the site with a slope of 10% or less, with the remaining portions of the site with slopes less than 40%
- Permeable surface, including unused existing lawns, and under/unused sites with impervious surfaces
- Direct, bright indirect, or moderate indirect light for at least 6 hours a day
- Water access, feasibility for the installation of new water access, or potential for rainwater capture
- Within reasonable walking distance from public transit
- Within reasonable distance from a vehicle drop-off area and reasonably accessible by a construction vehicle
- No streams or wetlands, including underground streams or gardens planned to be uncovered
- Rooftops and areas for vertical farms and gardens

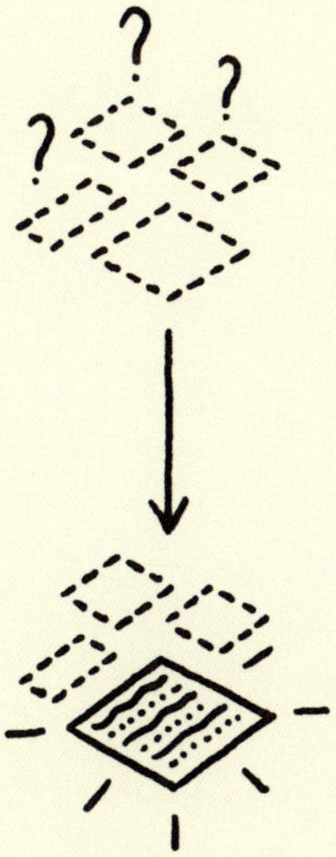

Individual agencies reviewed 120 sites and uncovered a total of 13 new parcels that were deemed available for gardening.

72 City of San Francisco. Executive Directive on Healthy and Sustainable Food 09-03. Summary Report. December 2010. pg. 7 and appendix F. Other cities that have gone through the process of searching for new land suitable for urban agriculture include Portland, Vancouver, and Oakland. See Mendes, W., Balmer, K., Kaethler, T., & Rhoads, A. (2008). Using Land Inventories to Plan for Urban Agriculture: Experiences from Portland and Vancouver. Journal of the American Planning Association. 74(4), 435-449.

1.3 Explore appropriate land tenure and garden preservation

Land tenure remains an issue for community gardeners on City-owned property. Several interviewees expressed concerns about the possibility of losing garden or farm space, which may influence the willingness and ability of practitioners to improve sites, their ability to obtain funding for projects, and the types of agricultural practices they are able to use on a particular site. Whether those fears are unwarranted, as some officials claimed, they persist among gardeners.

As described in the Urban Agriculture in NYC chapter, the community gardens that occupy City property have one-year licenses that can be renewed indefinitely, provided that the gardeners continue to use their space for gardening in compliance with all rules and regulations and the City does not wish to convert it to another use. To convert a garden site into another use, the City must go through a formal land use review process involving input from the local community board, the City Council, and Borough President, making the task time consuming and politically difficult. However complicated the process may be, the possibility remains for the City to develop the space.

The City should consider a range of license and lease terms for some community gardens and farms. For example, the Department of Parks & Recreation has entered into a long-term license with Added Value to enable the group to continue farming the Red Hook ball field it has occupied since 2003. Other cities have adopted different land tenure terms for gardens and farms. For instance, Baltimore[73] and Boston[74] recently issued requests for qualifications and proposals for urban farms that would be given five-year renewable leases, while the City of Oakland leases land to City Slicker Farms for 10 years. Longer-term leases enable farmers and gardeners to justify more substantial investments in their site's infrastructure, and may make it easier for them to obtain funding for their projects.

Another mechanism to provide long-term protection for urban agriculture sites is New York's nonprofit land trusts, which own and manage community gardens citywide. However, land trusts require funds to ensure that garden sites are properly maintained, and the city's existing land trusts would need to have additional revenue to add more gardens to their inventories. The Mayor's Office, Borough Presidents, or City Council, working with private funders, could help fund one or more land trusts to enable them to take on the ownership of additional sites.

1.4 Strengthen the role of the Food Policy Coordinator

To fully integrate urban agriculture into city government will require coordinating actions and policy-making across multiple agencies. One way to do this is to make the city's Food Policy

73 Request For Qualifications: Urban Agriculture in the City of Baltimore. RFQ Issued: March 25, 2011. Issued by: Baltimore City Department of Planning and Department of Housing and Community Development. Accessed at http://cleanergreener.highrockhosting2.com/uploads/files/Urban%20Agriculture%20RFQ%203.25.11.pdf on January 30, 2012.

74 http://www.cityofboston.gov/news/default.aspx?id=5188

Coordinator responsible for leading all efforts related to citywide urban agriculture policy. In 2008, a Mayoral Executive Order established the position of Food Policy Coordinator to improve New Yorkers' access to healthy food.[75] This position, which reports to the Deputy Mayor of Health and Human Services, includes the following responsibilities:

- Develop and coordinate initiatives to promote access to healthy food for all New Yorkers.
- Increase access to, and the use of, food support programs (such as the USDA's Supplemental Nutrition Assistance Program) for those who qualify for financial assistance to purchase food.
- Develop healthier City agency food standards for all meals or food supplies that are purchased, prepared, or served in agency programs.
- Ensure that meals and snacks served by city contractors and agencies meet those standards.
- Convene a Food Policy Task Force made up of representatives of various entities involved in the city's food system.[76]

One of the Food Policy Coordinator's critical functions is to foster collaboration across municipal agencies that typically are not considered responsible for food policy, but which control or manage elements of the food system such as land for urban farms and farmers markets. This mandate to coordinate the actions of multiple agencies makes the Food Policy Coordinator a logical choice to help agencies incorporate urban agriculture into their planning and operations. However, the executive order that established the Food Policy Coordinator position does not specifically mention urban agriculture. The City Council also could take the following steps to strengthen the position:

- Add urban agriculture to the list of food systems for which the Food Policy Coordinator is responsible.
- Establish the position through an amendment to the City Charter, making it a more permanent part of city government.
- Move the position to an agency with greater jurisdiction across multiple agencies, such as the Office of Operations.
- Provide the position with the adequate staff and funding to carry out the tasks of developing urban agriculture-related policy.

75 Executive Order 122 of 2008.

76 Executive Order 122 of 2008 states: "The Food Policy Coordinator shall convene a Food Policy Taskforce, which shall have representation including, but not limited to, the Department of Health and Mental Hygiene (the "Health Department"), Department of Education, the Human Resources Administration, the Office of the City Council Speaker, and the Council on the Environment of New York City." (emphasis added)

One of the first tasks of a newly strengthened Food Policy Coordinator could include working with other agencies to develop suitability criteria for urban agriculture on city-owned land. Additionally, the Food Policy Task Force should identify ways to streamline interactions among the agencies that provide services to or regulate urban farms and gardens. A top priority could be to develop a process—and information—to give gardeners and farmers a clearer sense of how to navigate various administrative requirements and access the services of diverse City agencies.

1.5 Increase the capacity of GreenThumb staff

One of the most significant obstacles to expanding urban agriculture in New York City is that much of the financial and logistical burden for supporting individual farms and gardens citywide falls on a single government entity—GreenThumb, a division within the Department of Parks & Recreation—that lacks adequate staff and funding. In order to allow a community group to start a new community garden, GreenThumb must interview the group to ensure that it has the capacity to manage the space, license the group, and oversee the use of the site to ensure that it remains well-tended and publicly accessible. In addition, GreenThumb provides services and materials, such as removing debris and weeds, soil remediation, new topsoil and compost, fencing, and a water supply. GreenThumb offers these services not only to the hundreds of gardens under its jurisdiction, but also to numerous nonprofit, commercial, and institutional farms citywide.

GreenThumb's current funding stream comes exclusively from Federal Community Development Block Grant funds.[77] These funds cover its staff costs for a director, several staff members, two to three outreach coordinators, and five employees who are involved in garden development, landscape restoration, and maintenance. These funds also pay for the program's budget, which includes everything from soil and lumber to gardening supplies, seeds and bulbs, vehicle use, information dissemination, training, and—budget permitting—sheds, hoop houses, and other garden enhancements. In fiscal year 2012, Green Thumb's OTPS (Other Than Personnel Services) budget was approximately $277,000.

GreenThumb is funded at a significantly lower level per garden than other leading municipal community garden programs, such as Seattle's P-Patch program. Moreover, the GreenThumb budget is scheduled to be reduced in the fiscal year 2013 financial plan. As the table at right illustrates, if the proposed fiscal year 2013 budget is adopted, GreenThumb will have a smaller budget than Seattle's community gardening program, despite having more than six times the number of gardens. It is unlikely that GreenThumb will be able to support additional gardens without an increase in funding.

77 Federal funding for GreenThumb comes from Community Development Block Grants, a federal program run by the U.S. Department of Housing and Urban Development to help cities stabilize and improve neighborhoods.

	Seattle P-Patch FY 2012 Expense Budget*	New York City Greenthumb FY 2012 Expense Budget**	New York City Greenthumb FY 2013 Financial Plan***
FY 2012 Expense Budget*	$ 669,071	$ 728,046	$ 655,454
Number of gardens	~ 75	~ 490	~ 490
Expenditure per garden	$ 8,921	$ 1,456	$ 1,338

* http://www.seattle.gov/financedepartment/12adoptedbudget/default.htm

** http://www.nyc.gov/html/omb/downloads/pdf/ss6_11.pdf

*** http://www.nyc.gov/html/omb/html/publications/finplan02_12.shtml

The City should ensure that GreenThumb's budget is adequate for its responsibilities for both management (tracking the gardens, renewing licenses, managing the delivery of supplies and services) and providing services and supplies. Allocating City funds to underwrite some or all of these costs, rather than relying on federal funds to do so, would enable existing farmers and gardeners to be more productive and facilitate the development of new urban agriculture projects.

In large organizations, including City government, what gets measured gets managed. One way to improve the management of GreenThumb gardens within City government is for the Department of Parks & Recreation to include these gardens in its evaluations of Parks facilities. For example, the Mayor's Management Report (MMR) tracks how well agencies perform their missions by outlining each agency's responsibilities and by reporting measurable indicators of success.[78] In addition, the report provides the public with information about how well particular agency tasks are being performed and how well the City is operating overall.

However, gardening or farming activities are not included in the MMR. As a result, the MMR does not include related indicators such as security issues at gardens, missing or broken fences, and compost provided by the Department of Sanitation, making it less likely for officials to commit funds to address these issues.[79] Including indicators for urban agriculture in a management tool like the MMR would make the needs of community gardens more visible, and therefore more likely to be considered as important as other agency functions—such as park maintenance—which are already tracked in the document. However, if community gardens are included in the MMR, the Department of Parks & Recreation should ensure that they are not penalized because they do not necessarily look or perform like other open spaces, such as manicured lawns or ball fields.

78 See http://www.nyc.gov/html/ops/html/data/mmr.shtml

79 The 2011 Mayor's Management Report does include the rest of the infrastructure managed by the Department of Parks & Recreation in the description of the scope of the agency's operations, from 1,800 parks to 600 comfort stations to 650,000 street trees. Community gardens, however, are omitted.

1.6 Establish an urban agriculture ombudsman

Because City agencies in New York do not view urban agriculture as part of their mission, staff at these agencies typically do not have the knowledge or expertise, or the administrative imperative, to address the needs of the urban agriculture community—access to compost or soil, learning how to set up a farm stand, or information about putting a greenhouse on a roof. As a recent study illustrated, community groups have to navigate substantial red tape to get permits for farmers markets.[80]

The City should designate one individual (or office) to receive public questions and concerns about urban agriculture, help farmers and gardeners navigate the bureaucracy, and serve as an urban agriculture liaison to other City agencies. This role could be assigned to the Food Policy Coordinator, the Office of the Public Advocate, an additional staff member assigned to GreenThumb, or designated staff within another City agency. This could initially be established as a pilot program to identify the issues that gardeners and farmers have, and to identify where there are real needs for agency coordination and collaboration.

80 Office of the Manhattan Borough President (2011) Red Tape, Green Vegetables: A Plan to Improve New York City's Regulations for Community-Based Farmers Markets. NYC: Manhattan Borough President's Office.

2. Integrate Urban Agriculture into Existing City Policies and Plans

Government agencies prepare plans for various aspects of the city's operations or physical development that address diverse issues, from affordable housing production to waste management.[81] Plans operate at different scales; rezoning plans, for instance, are often focused on neighborhoods, while transportation plans consider New York City within the broader region. Plans address the City's capital budget[82] and outline new and expanded public facilities required by City agencies.[83] In a well-planned city, each individual plan also addresses broader city goals and objectives, as well as topics addressed by multiple agencies such as health, equity, the environment, and economic development.

As it is in PlaNYC, urban agriculture should be treated as an issue that requires actions from multiple City agencies, which should integrate thinking about urban agriculture into their programs, plans, and long-range strategies. At a minimum, agency staff could assess how their plans and programs affect or could capture the multidimensional benefits of urban agriculture. This section outlines a number of opportunities for agencies to incorporate urban agriculture in their planning and operations, and potentially find cost-effective initiatives that can simultaneously achieve their goals, address other pressing citywide issues, and support farmers and gardeners.

2.1 Expand support for urban agriculture in the city's green infrastructure program

In 2010, the Department of Environmental Protection (DEP) released a stormwater management plan outlining the low-tech, landscape design techniques (collectively called "green infrastructure") that the City will use to slow and absorb stormwater runoff to reduce combined sewage overflows into our waterways.[84] DEP has committed to investing $187 million in green infrastructure over the next four years, and over $2.4 billion by 2030, to implement this plan.[85]

Although DEP's plan does not mention urban farms and gardens, the agency has funded several urban agriculture projects on private property as part of the green infrastructure program. Because urban agriculture provides multiple benefits, if DEP prioritized support

81 http://www.nyc.gov/html/dsny/html/swmp/swmp-4oct.shtml

82 New York City Charter, Chapter 10, section 228.

83 New York City Charter, Chapter 8, section 204.

84 http://www.nyc.gov/html/dep/html/stormwater/using_green_infra_to_manage_stormwater.shtml

85 This consists of $1.5 billion in public dollars (paid by water fees and state and federal funds) and $900 million in private investments, in addition to $2.9 billion in cost-effective conventional improvements.

for urban agriculture as a green infrastructure strategy, it could simultaneously tackle the combined sewer overflow problem and many other pressing citywide issues. For instance, in areas of the city such as the Bronx River watershed, new urban agriculture sites funded through the green infrastructure program would help reduce combined sewer overflows, while also providing fresh produce and new open space amenities in underserved neighborhoods, in addition to the other benefits associated with urban agriculture.

2.2 Establish a municipal soil conservation and distribution program

Soil is a critical resource for the city's gardens and farms, particularly those that use raised beds. The GreenThumb program purchases and delivers soil to registered school and GreenThumb gardens, but funds for soil purchases and deliveries are limited. In order to develop efficient, cost-effective strategies to address farmers' and gardeners' need for soil, the City could convene public and private partners to develop a citywide agricultural soils program. Participants could include City and State agencies such as Parks & Recreation, Sanitation, Environmental Protection and the New York State Department of Agriculture and Markets, the New York State Soil and Water Conservation District,[86] Cornell University's Cooperative Extension, and private contractors that currently deliver soil for the City.

The Need for Soil

One way to estimate the total amount of soil needed annually to support urban agriculture sites in New York City is to calculate how many sites exist citywide, how much land is used at those sites for growing food (as opposed to areas for community events, storage of tools, or other uses that do not require soil), and how much soil is needed for growing food.

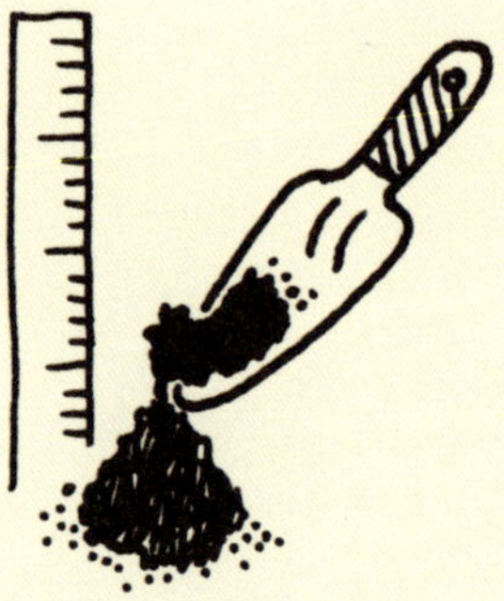

No one has yet established an accurate count of the total area of gardened and farmed land in New York City. Farming Concrete, in its 2010 report, noted that of 67 gardens surveyed there were 71,950 square feet of garden beds, averaging 1,074 square feet of bed space per community garden.[87] However, this survey captures only a fraction of the urban agriculture sites citywide. A more accurate citywide estimate would require an extensive survey of urban agriculture sites, which must include private backyard sites, school and public housing gardens, and commercial and community farms.

86 http://www.nys-soilandwater.org/about_us/what_we_do.html

87 Farming Concrete 2010 Report, p. 7. http://www.scribd.com/doc/53285030/FC-2010-Report

This program could conduct an initial analysis to determine the following:

- The total amount of land currently under production and the amount of soil needed to serve these sites

- The amount of soil needed over a ten-year period assuming the continued growth of urban agriculture. This could be roughly estimated in a number of ways, including assuming a certain number of additional acres of land are used for production each year for the next ten years

- The amount of soil currently provided by agency programs such as GreenThumb and other entities such as the city's botanic gardens and New York Restoration Project

- Options for storing soil in central locations and an efficient distribution infrastructure

2.3 Design a program to collect and compost organic matter, and distribute compost to gardens and farms

As part of its Compost Giveback program, the Department of Sanitation (Sanitation) began collecting residential leaves and yard waste in 1990 for composting at the former Freshkills landfill in Staten Island and Soundview Park in the Bronx, providing free compost for urban gardeners and farmers. In 2008, Sanitation suspended this program due to budget cuts, though it continues to compost leaf waste collected by private landscaping companies and the Parks Department, and helps to fund composting education and community-based composting projects.

The New York City Council recently passed two laws that pertain to citywide composting. Local Law 37 of 2010 requires Sanitation to resume the residential leaf and yard waste collection program by December 2012. The new collection program will run between March 1 and July 31, and from September 1 through November 30 of each year, a much longer period than the previous program, which collected leaves for only four weeks during the fall. In addition, the law requires Sanitation to collect leaf and yard waste from NYCHA properties, and to collect Christmas trees for two weeks instead of one week every January.

Local Law 42 of 2010 requires Sanitation, in conjunction with the Mayor's Office of Long-Term Planning and Sustainability, to issue a report by July 1, 2012, "recommending methods to expand the diversion of compostable waste from the city's waste stream." This report will also assess the feasibility of a curbside collection program

What is Compost?

Compost is the product of the decomposition of organic matter. When mixed with soil it provides nutrients and improves the soil structure. Finished compost can be produced in less than one year through properly managed decomposition of green plant materials and kitchen scraps, combined with dry organic material such as wood chips and paper.

COMPOST IN NYC

Compost is produced and distributed in many ways in New York City. The Department of Sanitation sponsors numerous compost-related programs, and scores of community gardens compost food scraps from local residents.

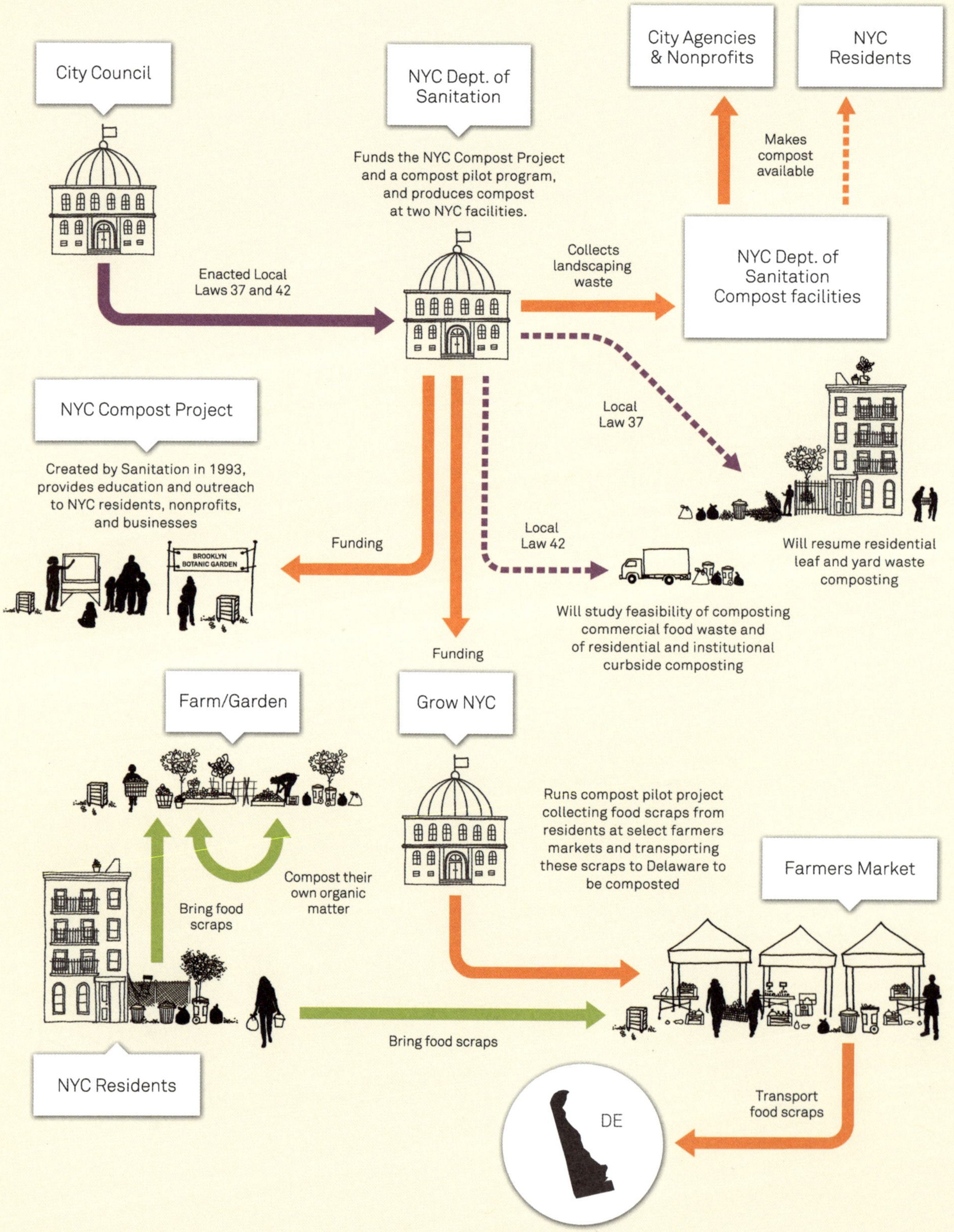

for household and institutional compostable waste, which includes food waste, paper products, and other organic matter. Local Law 42 also requires Sanitation and Long-Term Planning and Sustainability to provide a separate plan by 2014 to "study the viability of instituting a food waste composting program for the residential or commercial waste stream."

None of these regulations address the distribution of finished compost to gardens and farms. Although in the past, leaf and yard waste provided the primary material for Sanitation's now-discontinued Compost Giveback program, Local Law 37 does not require the agency to reinstate these giveaways. Furthermore, Local Law 42 does not require Sanitation's report on curbside compost collection and food-waste composting plans to address the needs of urban farmers and gardeners for compost.

In addition to conducting the analysis required by these laws, Sanitation should examine scenarios to make compost available for free or at low cost to the city's farms and gardens. Sanitation could consider how much compost the city's existing farms and gardens currently use and need, and how much would be required in future scenarios in which the city adds more farms and gardens. Sanitation also could evaluate the extent to which its compost production could meet current and future demand, and whether municipally-produced compost is a cost-effective way to supply farms and gardens citywide, and potentially throughout the metropolitan region.

2.4 Include urban agriculture in the City's review processes

Government agencies regularly assess the potential environmental and community impacts of their actions, such as the rezoning of a neighborhood, and the actions of private entities, such as constructing a new building, when their proposed projects require a City action. These review processes include the City Environmental Quality Review (CEQR) and the Uniform Land Use Review Procedure (ULURP).

These review processes currently do not require agencies to consider the effects of an action such as a new development or neighborhood rezoning on gardens or farms, or on access to healthy food.[88] **This analysis should be required by the Mayor's Office of Environmental Coordination, which assists agencies in carrying out environmental reviews and establishes parameters for the process, and the Department of City Planning, which oversees the ULURP process.[89]** For example, the review of a proposed building could consider if it would block sunlight from an adjacent community farm; potential mitigation measures could include reconfiguring the building, incorporating a new farm into the design of the proposed project, or creating a new farm somewhere in the project vicinity. Including this analysis of impacts on urban agricultural assets in these review processes would enable City agencies, developers, and the affected communities to determine what alternatives and mitigation measures, if any, are needed and feasible.

88 The GreenThumb licenses require that the City go through the ULURP process for garden sites they wish to develop.

89 The Manhattan Borough President proposed including food infrastructure in the City's environmental review process in 2009. See Planning for Healthy Neighborhoods: Include Food Infrastructure in the City's Environmental Review. A proposal by Manhattan Borough President Scott M. Stringer, June 2009. Accessed at http://mbpo.org/uploads/Food%20CEQR.pdf on March 16, 2012.

2.5 Incorporate agriculture into neighborhood planning

Urban agriculture requires the allocation of space, engagement of community members, and support from City agencies (from the police to sanitation workers) at a neighborhood scale. Citywide planning and policy-making for urban agriculture, which take a broad perspective and can address citywide issues such as the location of large-scale sites for composting, should be done in tandem with community-based planning, which addresses the local details of urban agriculture and taps the unique understanding that local residents bring to the table.

Examples of neighborhood-level planning issues related to urban agriculture include:

- Identifying sites for farmers markets and community gardens
- Addressing the compatibility of rooftop farms with neighborhood characteristics
- Determining the desirability of integrating farms into new residential buildings
- Locating greenhouses and identifying land for food processing facilities
- Allocating capital dollars for municipal food processing facilities
- Considering the use of parkland for growing and selling food

Two entities—the Department of City Planning (DCP) and Community Boards—routinely handle issues that affect urban agriculture. For instance, Community Boards help shape the City's capital and operating budgets, which can be directed to support urban farms and gardens.[46] DCP reviews neighborhood plans and changes to the City's zoning resolution, which, among other things, regulates where different uses (such as businesses, apartment buildings, factories, and parks) can be located.

What is the City Environmental Quality Review process?

The City Environmental Quality Review (CEQR) is the process by which City agencies review proposed actions for their environmental impacts. For instance, for a proposed new building that requires a zoning variance, the CEQR process would consider how the building might impact the existing historic character of the neighborhood, its open spaces, and natural resources; whether the existing transportation network is adequate to handle any additional traffic that might result when the building opens, and the shadows the building might cast on adjacent properties.

Where substantial negative impacts are likely, CEQR generally requires an Environmental Impact Statement (EIS) to examine those impacts in detail and propose potential mitigation strategies and alternatives. The environmental review process is no guarantee that projects will be improved as a result of the analysis. Nevertheless, an EIS is often the only publicly available sources of detailed data and analysis to enable communities and decision makers to understand, comment on, and avoid the anticipated negative consequences of new projects and programs.

What is ULURP?

The Uniform Land Use Review Procedure (ULURP) ensures that New Yorkers can review proposed land use changes, from the sale of City-owned property to the designation of urban renewal areas, before land is conveyed or designated. ULURP mandates that the affected community board and the Borough President consider and vote on land use changes before they can be approved by the City Planning Commission and—for some actions — by the City Council. This process governs decisions by the City to sell property on which community gardens exist, or to purchase land for new gardens. (No such process exists for gardens located on private, state, or federal land in New York City.)

Community Boards and DCP should use their authority proactively to assess the physical and resource needs for urban agriculture within their community and incorporate urban agriculture into neighborhood-level planning processes. These include: rezonings, 197-a plans, the Uniform Land Use Review Process (ULURP), the Fair Share Process, and the budget-setting process.

Rezoning

Since 2002, DCP has completed 115 neighborhood rezonings (changes to the City's zoning resolution), none of which explicitly consider the impacts on the city's farms and gardens. DCP could revise the rezoning process to include assessing the impacts that land use changes and the new development might have on existing farms and gardens. It also could explore ways to provide space for new urban farms and gardens within rezoned neighborhoods.

197-a plans

Section 197-a of the City Charter authorizes the production of plans for specific areas within the city. Thirteen 197-a plans have been adopted citywide to date. DCP could encourage 197-a plans to incorporate urban agriculture. For instance, a 2007 197-a plan approved by DCP for Manhattanville and Morningside Heights states that one of the primary goals is to "build on the strong social, economic, and cultural base of the district

Community Boards

Community Boards are a system of local representative bodies in New York City. Each of the city's 59 Community Districts is represented by a Community Board, and each Board is made up of 50 unsalaried members. Community Boards are responsible for shaping municipal priorities in land use, advising on the services provided by municipal agencies, recommending agency budget expenditures, and evaluating the environmental impacts of proposed projects.

The composition of community boards may not be representative of the neighborhoods they comprise, resulting in policy positions that do not reflect all the constituents of that community. Because they are composed of citizens who are unfamiliar with urban agriculture, community boards generally require technical assistance to engage effectively in food systems and urban agriculture planning. However, despite these limitations, Community Boards remain an important vehicle for community participation in neighborhood planning.

through a sustainable agenda that would reinforce and reinvigorate the ethnically and culturally diverse neighborhood."The plan recommends facilitating new green areas, including locations for new "landscaping and plazas on city-owned properties," and studying the feasibility of establishing an additional farmers market within the district. These kinds of proposals could easily incorporate an assessment of existing and potential urban agriculture sites.

In those communities developing or revising a 197-a plan, Community Boards with vacant land could identify city-owned parcels that the board deems appropriate for small-scale urban agriculture for food production through the 197-a process.

ULURP

A Community Board could work to ensure that neighborhood-rezoning projects include adequate space for urban agriculture. Through the ULURP process, boards could provide extra scrutiny over proposals for developing community gardens and farms, or for development projects that jeopardize garden or farm sites.

Tools to Assist in Neighborhood-Level Planning

In 2011, members of the nonprofit organization Food Systems Network NYC launched a project called FoodAction NYC to help members of the public to work with community boards on food planning projects.[90] FoodAction will produce an online toolkit, along with information on the city's food system that can be mapped to help community boards to engage in food systems planning.[91]

Fair Share Process

The Fair Share Process is used to equitably allocate sites for municipal facilities.[92] Each year, the Mayor's Office publishes a document called the Citywide Statement of Needs, a compendium of proposed additions, closures, expansions, or other plans for changes at city facilities for the coming two years. The Statement of Needs is based in part on the needs assessed by each individual community board through a budget-setting process (described below). While much of the original impetus for this process was to prevent the unfair clustering of potentially harmful facilities such as waste treatment plants, it also could be used to ensure that urban agriculture–related facilities such as compost distribution sites or new farms and gardens are being distributed fairly throughout the city.

Budget Priorities

Community Boards also submit budget proposals based on their district's needs to the Office of Management and Budget (OMB) as part of the process to determine citywide budgets. This process allows Community Boards to make both capital and operating budget requests, such as capital funds to construct lighting and other infrastructure to support outdoor farmers markets, or operating funds for a specific community farm. Community Boards should work with farmers

90 http://www.foodsystemsnyc.org/articles/flour-goes-local?page=4

91 http://www.oasisnyc.net

92 http://www.nyc.gov/html/dcp/html/pub/fsguide.shtml

and gardeners, support organizations, and GreenThumb to make budget requests that address the needs of farms and gardens in their district. It should also be noted, however, that the Community Board process is not sufficient to gauge the need for capital funds for urban agriculture, especially in low-income communities that may have more pressing needs besides urban agriculture.

Neighborhood Urban Agriculture Planning in Chicago and Vancouver

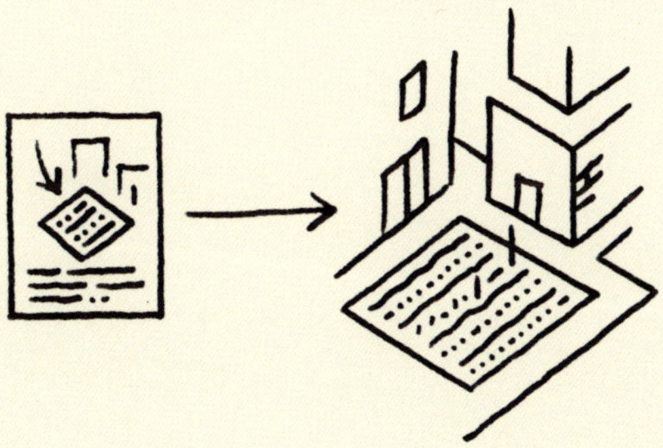

Chicago's city planning department has incorporated urban agriculture into several neighborhood plans. In the Englewood community, on the city's south side, city officials established the Greater Englewood Urban Agriculture Task Force, which includes Growing Home (a community-based urban agriculture organization), other community groups, local financial institutions, the local community college, and Englewood residents. The Task Force's goal is to turn Greater Englewood into a "food destination" by creating a large number of urban farms, developing and promoting farmer training, planning for business and infrastructure development that supports new and existing food entrepreneurs and urban farmers, and offering food education through the local community college.[93]

Vancouver's Greenest City Action Plan (GCAP) addresses the regional, national, and global dimensions of the food system, yet focuses on neighborhood-level initiatives[94] The Plan calls for the development of neighborhood food infrastructure, including food hubs (food processing and distribution facilities to connect rural farmers to urban consumers), community kitchens, markets, gardens, and even community root cellars for food storage and community bread ovens. The city has provided small grants to a neighborhood-based network, called Village Vancouver, to develop more localized food systems and reduce energy consumption, food waste and packaging, and to achieve other environmental goals.[95] Specific projects include asset mapping in neighborhoods to identify community resources for urban farming, including land and skills, and training programs to enable members to start seed-saving collectives, food co-ops, beekeeping organizations, and other projects.

93 Greater Englewood Urban Agriculture Task Force (GEUA): Progress, Accomplishments and Priorities for the Coming Year. no date.

94 City of Vancouver. Administrative Report No. A-8, June 8, 2010. Accessed at vancouver.ca/ctyclerk/cclerk/20100622/documents/a8.pdf

95 http://www.villagevancouver.ca/

3. Identify Innovative Opportunities to Build Urban Agriculture into the Cityscape

At a smaller scale than neighborhood-level plans, government agencies routinely plan for individual buildings or parcels of land. Because urban agriculture can thrive in very small areas in many different forms, New Yorkers have already reclaimed many different kinds of spaces citywide, from rooftops to stalled development sites, using everything from milk crates to kiddie pools to grow food. Government agencies have many opportunities to facilitate this activity and encourage the creation of new farms and gardens by modifying existing regulations and procedures for developing City-owned land.

3.1 Support project-level urban agriculture planning and design

The NYC Department of Housing Preservation and Development (HPD) supported the development of a 202-apartment affordable housing project in the Bronx called Via Verde that includes rooftop gardens designed to encourage multiple uses, such as fruit and vegetable cultivation, passive recreation, and social gathering, while also providing the benefits of stormwater control and building insulation. The developer of Via Verde, Jonathan Rose Companies, was selected from a competitive Request for Proposals (RFP) process. Although the RFP did not specify an urban agriculture feature, it did require respondents to consider incorporating access to nutritious food, a fitness theme, and places for social gathering into the development.[96]

Urban Agriculture at Via Verde

Spinach, kale, collards, broccoli, cabbage, onions, and peas are growing at Via Verde in 11 raised garden beds, totaling approximately 1,541 square feet, while the building's fourth-floor terraced roof features apple trees in a dozen large rooftop planter boxes. In the 20th-floor community room of the Via Verde tower, GrowNYC staff will provide monthly workshops on gardening, composting, and healthy eating for Via Verde residents, using the facility's kitchen to prepare meals using produce from the gardens. Residents will be encouraged to participate in the weekly rooftop garden and harvest work days.

96 City of New York. Department of Housing Preservation and Development. (2006) New Housing New York Legacy Project Request for Proposals. Issue Date: June 12, 2006. Accessed at http://www.aiany.org/NHNY/rfp/index.php on January 20, 2012.

Encouraging Urban Agriculture in Affordable Housing Development

The Enterprise Foundation's Green Communities[97] Certification program uses a point-based evaluation tool for affordable housing projects that is similar in approach to the Leadership in Energy and Environmental Design (LEED) standards for buildings. Developers get points for including various features to improve public health and food access – urban agriculture is one of those features. One point can be earned by incorporating one of the following strategies for increasing access to healthy foods:[98]

- Building neighborhood farms and gardens into the project (including permanent growing space and related infrastructure)
- Connecting the project to established community gardens within a half-mile walk
- Offering the delivery of community-supported agriculture (CSA) program shares for residents, project staff, and surrounding community
- Locating the project within a half-mile of an existing or planned farmers market

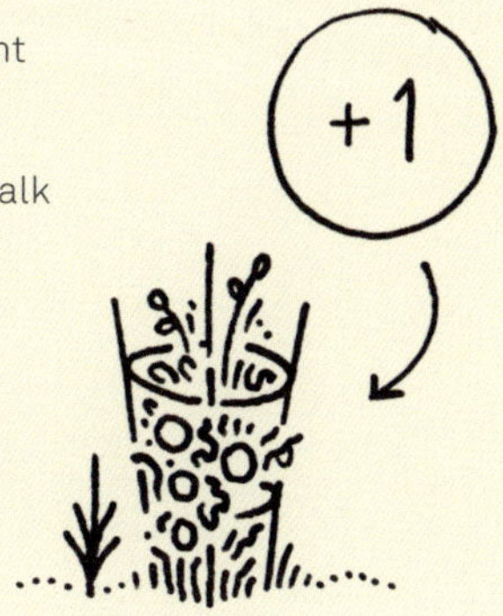

HPD and other agencies should encourage new development projects to include growing spaces by specifying the incorporation of urban agriculture and food access, more broadly as part of the project program and design guidelines. There are precedents for doing so. State policies, for instance, encourage developers of affordable housing to use green building techniques. And at the city level, HPD has incorporated the Enterprise Green Communities Certification as a threshold requirement for obtaining Low-Income Housing Tax Credits.[99]

3.2 Encourage rooftop urban agriculture

Urban farmers and gardeners in dense cities increasingly are interested in using rooftops as growing spaces. New York has become a leader in developing this form of urban farming, with a wide array of rooftop agricultural projects. Nonprofits have used roof spaces for therapeutic gardens, housing developers have built rooftop gardens and greenhouses as amenities for new residential buildings, and grocers and restaurants are sourcing fresh produce from their roofs. On an administrative building on Randall's Island, the Department of Parks & Recreation is testing different kinds of green roof strategies, including vegetable plots, that in the future may be installed on recreation centers citywide. In addition to providing food and open space, these projects collect stormwater and can reduce the buildings' heating and cooling costs.

97 http://www.enterprisecommunity.com/solutions-and-innovation/enterprise-green-communities/resources/policy-support-n-r#new-york

98 www.enterprisecommunity.com/solutions-and-innovation/enterprise-green-communities/certification

99 Criterion number 2.12

Rooftop Greenhouses

To qualify for the new exemptions approved by DCP, rooftop greenhouses must:

- Not be on buildings that contain residences or other uses with sleeping accommodations.[100] DCP believes that residential building owners will turn rooftop greenhouses into additional living space instead of growing space.
- Only be used to grow plants (or if they are accessory to a community facility, are used primarily for plant cultivation)
- Not exceed the building height limit by more than 25 feet
- Have roofs and walls that have at least 70 percent transparent material (not counting for accessory office or storage space, which may take up no more than 20% of the floor space and have solid walls and roofs)
- Be set back from the perimeter wall by at least 6 feet all around if the greenhouse exceeds height limits
- Incorporate a rainwater collection and reuse system to reduce the demand on the potable water supply and minimize stormwater runoff

New York has up to 2,700 acres of flat rooftops with the potential to serve as growing spaces—that's an area more than three times the size of Central Park.[101] New York City planners and policy makers have been trying to make it easier to install rooftop greenhouses by changing regulations on the height and size of structures on top of existing buildings. Local Law 49 of 2011 amends the building code by adding greenhouses to a list of other rooftop structures, such as water tanks and air conditioning equipment, that do not count toward height restrictions. The Department of City Planning approved a zoning text amendment that excludes rooftop greenhouses atop commercial buildings from regulations that limit the bulk of the building.[102]

City agencies should continue to look for ways to encourage rooftop agriculture. The Department of Environmental Protection could expand its support for rooftop farms as part of its program to manage stormwater. Through its RFPs, HPD could incentivize housing developers to include greenhouses in new buildings. City Council members and Borough Presidents could use their discretionary budgets to help fund rooftop greenhouses.

100 Greenhouses are allowed atop residential buildings if they stay within the FAR and height limits for the site.

101 Ackerman, K., op cit.

102 The text amendments exclude greenhouses from the calculation of the floor area ratio (FAR) of the building. See http://www.nyc.gov/html/dcp/html/greenbuildings/index.shtml

3.3 Support "interim use" urban farm projects

Numerous urban agriculture projects are designed to occupy spaces temporarily. On East 29th Street in Manhattan, for instance, GrowNYC is managing the 15,000 square foot Riverpark Farm on the site of a stalled real estate development. Vegetables have been planted in 3,400 milk crates, which can be moved when construction resumes.[103] Other municipalities have incentivized developers to use stalled sites for urban agriculture and other "green" uses. **The Department of Buildings and the Department of City Planning could follow this lead to encourage temporary urban agriculture projects on the more than 600 stalled development sites citywide.**

Riverpark Farm, Manhattan

Urban Agriculture on Stalled Development Sites

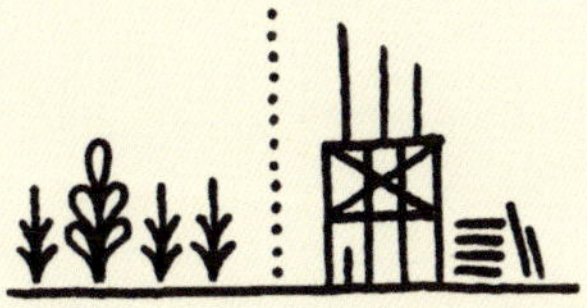

In San Francisco, city officials created an innovative strategy, called a Green Developer Agreement (GDA), to encourage temporary green uses of sites where development is stalled.[104] This binding agreement allows developers to preserve their development approvals for a five- to eight-year period as long as the site is used for a "green" purpose, including urban agriculture. The agreement ensures that the interim "green" use remains until construction begins, while protecting developers from losing control over the site in case the interim use proves so popular that residents attempt to scuttle the original approved development. In 2010 a nonprofit organization opened an interim-use community farm project on a vacant parcel near the city's downtown.

103 Collins, G. August 2, 2011. Tom Colicchio's Secret Farm, Right Next to the F.D.R. Drive. The New York Times.

104 San Francisco Planning Department. (2010) Executive Summary: Draft Green Development Agreement Legislation. Accessed at http://sf-planning.org/ftp/files/Commission/CPCPackets/Commission%20Packet%20for%20Green%20Development%20Agreement%20Legislation%20for%20February%2025%202010%20Hearing.pdf

Forest Houses

In December 2010, the New York City Housing Authority sold a parcel of land at the Bronx public housing project Forest Houses to Blue Sea Development to facilitate the construction of 124 units of affordable housing. When it opens, the development will feature a 10,000-square-foot hydroponic rooftop greenhouse operated by a private firm, Sky Vegetables, that will grow produce on a commercial basis for the surrounding community.[103] The initiative to put a greenhouse atop the development was entirely Blue Sea's, though the support of local elected officials was critical.

The project overcame three obstacles that developers of other building-integrated greenhouse projects may face. First, City agencies and tax credit investors unfamiliar with building-integrated agriculture were unwilling to allow the anticipated revenues from the urban agriculture operation to be incorporated into the building's pro-forma, making it effectively impossible to finance the costs of the greenhouse through anticipated rental revenue from its operations. In the end, Blue Sea was able to secure funds from the Bronx Borough President's Office and the City Council to cover the greenhouse purchase and installation, but was precluded from charging rent for the space.

Second, the developer had to address a zoning issue. Blue Sea had to get a determination from the Department of City Planning that a greenhouse is an agricultural use that counts as a community facility, which is permitted atop residential uses. The greenhouse will sell produce within the neighborhood, reaching people within the community most in need of greater access to fresh food, and distribution of the produce will be done off-site through a CSA and other distribution channels.

Finally, the Department of Buildings initially required the greenhouse to meet the same energy code standards that apply to the whole building, a standard that freestanding greenhouses are not required to meet. Negotiating with the Department of Buildings over the energy code issue delayed the project by several months.

3.4 Encourage gardening in small spaces

Given that New Yorkers grow herbs and vegetables on windowsills and balconies, and in containers in a variety of small spaces, even tiny vacant parcels can be used for food production. Some farmers and gardeners have even found ways to aggregate small, dispersed garden plots, like backyards, to grow enough produce to run a CSA.[106]

Other small public spaces can be used to grow food as well. In Seattle, for example, residents are allowed to grow edible plants in what are known as "parking strips," the space between the curb and sidewalk found in residential neighborhoods.[107] The Vancouver environmental education organization City Farmer has a demonstration curbside vegetable garden to illustrate the possibilities of growing food in such a space. Portland's Bureau of Transportation also allows edible landscaping in parking strips. New York City amend-

105 U.S. Department of Housing and Urban Development Office of Public and Indian Housing. (2011). Final PHA Plan. Annual Plan for Fiscal Year 2012. New York City Housing Authority. October 18, 2011. pg. 7. Accessed at www.nyc.gov/html/nycha/downloads/pdf/FY2012-AnnualPlan.pdf on 1-20-12.

106 BK Farmyards. Accessed on 3/10/12: http://bkfarmyards.blogspot.com/p/foxtrot-farmyard.html

107 Seattle, City of, Department of Transportation. (2011) "Department of Transportation Client Assistance Memo 2305: Gardening in Planting Strips." January 1. Available at: www.seattle.gov/transportation/stuse_garden.htm (last accessed April 26, 2011). (Clarifies that residents may plant raised-bed gardens in the strip of the public right-of-way between the sidewalk and the curb.)

ed its zoning text in 2009 to require landscaped parking strips but allows only grass or groundcover to be planted.[108] **New York City could modify its zoning to allow growing vegetables in parking strips to encourage more residents to become active stewards of what is typically a neglected part of the streetscape.**

3.5 Strengthen infrastructure for food distribution and production

The city's farmers and gardeners could benefit from shared large-scale infrastructure and centralized facilities. Other municipalities have created facilities known as "food hubs" that combine the following:

- Shared trucks for farms and gardens to pick up material like soil and compost, and to deliver produce to farmers markets, vendors, or other retail outlets
- Refrigeration equipment to reduce the waste of unsold produce at farmers markets and enable weekly harvests to be sold on more than one market day
- Processing facilities to add value to and preserve the produce grown on farms and in gardens
- Warehouses to aggregate produce from multiple gardens and farms, and potentially serve as a distribution point to multiple markets

Food Hubs

The Stop Community Food Centre is a community-based facility in Toronto that includes an 8,000-square-foot garden, a greenhouse, a "global roots" garden to demonstrate culturally specific foods, and many other food-related projects. For example, the Stop runs a farmers market, a café, and a food share program that buys and redistributes produce through farm stands located across the city. The organization also has facilities to teach cooking and nutrition, an after-school program and summer camp, and other educational programs.

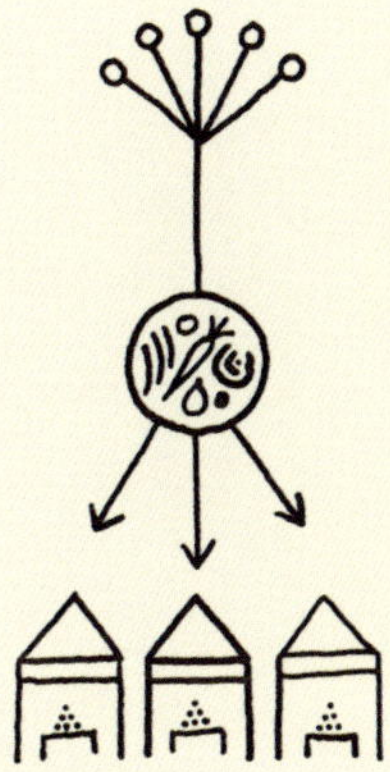

Detroit's Eastern Market is the city's wholesale food distribution facility, with a six-block farmers market for 250 local vendors surrounded by food distributors and other food establishments. Eastern Market has a demonstration urban farm, an after hours wholesale market, and special food-related public programming to draw customers to the market.

108 See http://www.nyc.gov/html/dcp/html/street_tree_planting/index.shtml

The Economic Development Corporation could assess the costs and benefits of creating one or more such facilities to promote microenterprise opportunities. This assessment could include the following:

- The management structure, design, amenities, costs, and benefits of precedent projects, such as the Stop in Toronto, Detroit's Eastern Market, and La Marqueta, a kitchen incubator in East Harlem

- Demand and capacity among local farmers and gardeners for facilities to produce value-added items, such as jams and salsas, to store food from each harvest not given away or sold at farmers markets, and to share vehicles for distribution of food or to pick up materials such as soil or compost

- Opportunities and barriers to sharing existing, underused facilities, such as institutional kitchens in schools, universities, and community centers, and cold storage space

- Criteria for locating new microenterprise facilities, including proximity to farms, gardens, commercial districts, and socioeconomic indicators of need

3.5.1 Procurement

To stimulate the development of regional farms and supply City agencies with fresh food, Local Law 50 of 2011 requires the City Procurement Officer to develop guidelines for agencies to maximize the purchase of food produced in New York State, and to submit an annual report to the City Council each October 1 detailing agency efforts to do so. This legislation was crafted with the expectation that it would apply to regional farms outside New York City seeking an opportunity to compete in the City's purchasing process. However, **these guidelines also could direct agencies to purchase some produce from urban farms and gardens**, provided that the costs of doing so are equal to the costs to the City of produce procured from any of its current sources.

4. Address Disparities in New York City's Urban Agriculture Community

In many ways, disparities within the city's urban agriculture community reflect larger societal inequalities. Numerous philanthropic, nonprofit, and municipal programs attempt to address these disparities, targeting assistance, resources and funding to farmers and gardeners in low-income neighborhoods. Yet much more could be done to ensure that opportunities are equitably available to all farmers and gardeners in New York City, and that decision-making processes are transparent and engage a broad range of urban agriculture stakeholders.

4.1 Increase access to information about available resources

Numerous interviewees said the process of securing resources from the City was unpredictable, in part because not all farmers and gardeners have the time or resources to establish relationships with municipal agencies. In order to ensure that all farmers and gardeners have full access to information about funding opportunities and agency programs that support their missions, City agencies should establish clearer guidelines for how they work with farmers and gardeners, and make this information publicly available and easily accessible.

These guidelines could address the following questions:

- What materials are available for gardens, and what is the procedure for ordering and receiving them?
- What decision criteria will be used to allocate scarce resources across many gardens and farms?
- What are the needs of the urban agriculture sector for capital improvements, and how can those be addressed equitably across the city?
- How do agencies decide with whom to partner on projects and programming, and therefore how can organizations that have not historically had access to those resources compete for them?

Another step would be to publish a single list of all of the sources of public funding which are distributed to gardens and farms each fiscal year. Such information would reveal funding disparities, could help determine which organizations need help with capacity building to be able to access public funds, and would identify opportunities and priority areas for funding from private funders. The City Council already publishes a list of all Council Member grants to organizations (whether or not they are urban agriculture-related) within their own districts. In order to have a clearer picture of where and how much funding is being disbursed for urban agriculture overall, a singular and specific list would need to be compiled with data from individual agency budgets, the City Council, and the five Borough Presidents.

4.2 Support capacity building for underserved urban agriculture groups

Government agencies should support capacity building among urban agriculture organizations that historically have not received substantial public or private funds, particularly those focused on social justice and/or led by people of color. This support could include training and assistance on bookkeeping, fundraising and grant writing, organizational management, data collection, program evaluation, and business practices. For instance, one funder has piloted a targeted funding stream to support organizational and leadership development among food organizations led by people of color; this could provide a model for other funders, support organizations, or government agencies. This type of capacity-building support would help groups compete for funding, manage their farms and gardens more efficiently, and potentially identify new sources of revenue.

The City also could provide additional funding to support organizations that already provide technical assistance to farmers and gardeners for food production so that they might expand their portfolios to include business planning, evaluation and service design, making sure that this assistance targets historically underserved groups.

Agencies could facilitate strategic partnerships between funders and support organizations to broaden their outreach. Large foundations without detailed knowledge of the city's urban farms and gardens can partner with local supporting organizations that may be better positioned to reach a broad range of applicants. For example, the nonprofit Green Guerillas worked with foundations to obtain sizeable grants, and then regranted funds to gardening or farming programs for the purchase of relatively low-cost supplies, such as garden tools.

Agency programs should also encourage urban farms and gardens to partner with each other and other organizations on projects and funding applications. This may create synergies and economies of scale among the urban agriculture community and encourage sharing of knowledge and resources, while assisting traditionally underfunded groups in securing grants. For instance, joint funding could be offered to a coalition of farms and gardens to share tools or to hire a single youth educator who could rotate among different sites.

4.3 Provide resources and assistance with community development and outreach

Urban gardeners and farmers are integrally connected to their neighborhoods, yet a number of interviewees indicated that they wanted to engage more people in the surrounding community. They also felt that City agencies did not always sufficiently value local residents' networks and expertise, missing opportunities to include farmers and gardeners in leading neighborhood-based education and outreach programs.

GreenThumb and support organizations such as Green Guerillas currently offer assistance on community organizing and public programming, but have limited funds to do so. The City could provide additional financial resources to these groups, enabling farmers and gardeners to expand their outreach and recruit more local residents to grow food and participate in programs, while avoiding practices which negatively affect their neighbors' quality of life. This assistance could also accommodate the growing number of New Yorkers who are interested in starting new gardens and farms.

4.4 Establish equitable and transparent participation in policy-making

Numerous interviewees said they felt disconnected from the policy-making process or were included only nominally in policy decisions. Interviewees also noted that City officials could benefit from collaborating more with farmers and gardeners, and their existing community networks, to implement agency programs.

City agencies should develop guidelines for public participation in policy-making about urban agriculture, including systems for ensuring representation of NYC's diverse communities and neighborhoods in these processes. For instance, **the City should establish, by law, a citywide urban agriculture task force.** This group, made up of farmers and gardeners, intermediaries, and City officials, would review the programs, policies, and budgets that affect urban agriculture and advise policy makers on strategies to strengthen and expand urban agriculture. There is ample precedent in New York City government for such advisory bodies which have addressed issues ranging from sustainability and waterfront management to solid waste disposal and recycling.

A task force could have an immediate impact on urban agriculture in New York City, helping to establish criteria for new sites for farms and gardens, advising the Office of Long-Term Planning and Sustainability on the urban agriculture–related commitments in PlaNYC, and shaping the scope of any future citywide urban agriculture plan. To ensure fair representation of the city's farmers and gardeners on the advisory board, and thereby address one aspect of the concerns about race- and class-based disparities in the urban agriculture system, the process for selecting advisory board members should itself be developed in consultation with a diverse range of key urban agriculture stakeholders.

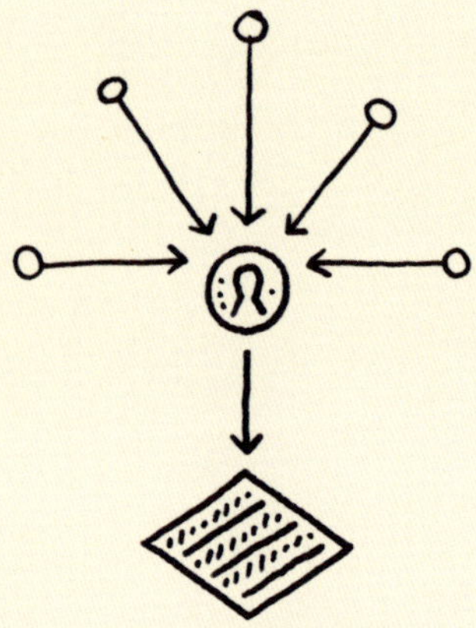

Participatory budgeting

One precedent for democratizing the budget process in New York City is participatory budgeting, a process through which community members decide how to use part of a public budget. Four City Council members used participatory budgeting to allocate a portion of their discretionary funds in Manhattan, the Bronx, Brooklyn, and Queens, a process which began in late 2011 and was completed in April 2012. Through participatory budgeting processes an elected official convenes a broad range of stakeholders to discuss their priorities and how much specific budget items would cost. Participating stakeholders then vote for how the budget should be allocated.[109]

Participatory budgeting is new to New York City and is therefore in a preliminary stage of development. Although this process may need to evolve to address shortcomings, including the possibilities for unfair distribution of a public budget due to who is (and who is not) able to participate, and what those individuals' priorities might be, participatory budgeting offers one potential strategy for increasing community engagement in budget processes.

4.5 Engage the urban agriculture community in the budget process

New York City's budget process includes several opportunities at the neighborhood, borough, and citywide level for individuals and organizations to recommend increased funding for urban agriculture, but relatively few New Yorkers participate in these processes. An important role for city and elected officials would be to actively engage urban farmers and gardeners in the budget planning and adoption process, helping them to provide input to the budget process from the community board level up to the Mayor's preliminary and Executive budget.

4.6 Commit to improving agency-level capacities to address race- and class- based disparities

As noted above, race- and class-based disparities stem not only from individual and intentional biases, but also unintended and institutionalized patterns. There are numerous community-based and larger organizations in New York with experience addressing race- and class-based inequities. **The City should enlist the assistance of these groups to provide advice about how its agencies can address disparities in New York City's urban agriculture system in both the short and long term.** Specifically, City agency processes could be informed by consultation with citywide or national organizations and grassroots community groups whose urban agriculture activities center on food justice, community empowerment, and anti-racism.

109 Accessed at http://pbnyc.org/ on February 24, 2012.

5. Urban Agriculture Grantmaking

The philanthropic community has a significant opportunity to increase the capacity of urban agriculture organizations, and influence the scale, scope, and focus of their practices. At a time when vital municipal services are regularly under threat of budget cuts, foundations can not only fund specific farms and gardens but also leverage public investment in a wide range of initiatives to support urban agriculture.

The many foundations currently funding urban agriculture in New York City have diverse agendas and priorities. This section focuses on four broad areas that the philanthropic sector is perhaps uniquely positioned to address.

5.1 Equalize grant opportunities throughout the urban agriculture community

The philanthropic community can help to ensure that groups led by people of color and groups that have limited fundraising capacity have access to grant opportunities. This involves increasing the capacity of minority-led groups, and organizations that need board development and other organizational help, to compete for funds. Funders could identify geographic and resource gaps in the current distribution of philanthropic dollars, and encourage groups to partner on grant applications, pooling their resources rather than competing for funding.

5.2 Explore a sustainable funding source for urban agriculture

"All the not-for-profits have a lot of money problems right now... Demand [for urban agriculture] is huge, the more it gets promoted by the media, the Mayor, everybody else... but there's not people writing huge grants for [these organizations]."

– Government Official

Obtaining funds and resources remains unpredictable for farms and gardens, agency programs such as GreenThumb, and support organizations. At the same time, many New Yorkers and private businesses have shown tremendous interest in supporting the city's urban agriculture movement. In partnership with farmers, gardeners, and support organizations, the philanthropic community could explore strategies to engage individual donors, corporations, and federal agencies (such as the Department of Agriculture) as potential sources of funding for urban agriculture and other resources such as soil and tools.

One funder interviewed for the project is involved in the creation of a funder affinity group to foster complementary and potentially collaborative funding for food- and health-related issues. A similar group could be formed to develop a shared funding strategy for the urban agriculture community. This group also could build the capacity of advocates to influence federal policy, enabling community-based urban agriculture organizations to link with national policy advocates.

5.3 Provide support for more networking among farmers

The city's farmers and gardeners have several opportunities to attend large-scale events attended by hundreds of people in the urban agriculture community. For instance, GreenThumb hosts Grow Together, an annual, low-cost educational and networking event focused on food production, and a number of organizations such as Just Food, the NYC Community Gardening Coalition, Black Urban Growers, and the Brooklyn Food Coalition also host well-attended conferences. Many urban farmers and gardeners also participate in regional events, such as conferences on sustainable agriculture hosted by the Northeast Organic Farming Association.

Farmers and gardeners would benefit from two other kinds of opportunities to meet each other and network: small groups that gather farmers and gardeners from a few adjacent neighborhoods, and meetings that connect urban and rural agriculture. For instance, Green Guerillas has helped organize regular meetings in Central Brooklyn for community gardeners to share information about how to recruit local residents to join gardens, apply for funding, and other best practices. Funders could support more of these types of neighborhood-level meetings.

Urban farmers and gardeners of all types cited the value of their ties to regional farmers, while noting that it is difficult for urban growers to make connections with their rural counterparts. Commercial farmers, in particular, said they welcomed more opportunities to share information about issues such as growing and marketing techniques and business planning with regional commercial farmers. A number of organizations and agencies could help urban and rural growers exchange information with each other through urban-rural growers' meetings or other networking strategies. These groups include Cornell Cooperative Extension, the NYS Department of Agriculture and Markets, American Farmland Trust, and GrowNYC, which has a network of regional farmers who sell at its Greenmarkets and participate in other programs.

Conclusion

Among a number of people interviewed for *Five Borough Farm*, there was a perception that urban agriculture can thrive without much municipal support. Many farmers and gardeners are extremely self-reliant and resourceful entrepreneurs, but as this chapter illustrates, the policies and practices of more than a dozen City agencies already affect urban agriculture in hundreds of different ways, even if they are not explicitly focused on food production. How the Sanitation department picks up and manages organic waste, whether or not an agency makes productive use of the land under its jurisdiction, how a developer designs a new affordable housing project, who gets public funds to create green space in neighborhoods plagued by stormwater discharges – these are all policy decisions made on a regular basis that could maximize benefits to farmers and gardeners, their communities, and the city as a whole.

This chapter outlines a plan of action to grow urban agriculture in all five boroughs, including strategies to more thoroughly integrate farms and gardens into the cityscape, and programs, policies, and practices that include urban agriculture in the day-to-day decision-making of a broad range of City agencies. Over the next year, in collaboration with the Department of Parks & Recreation, the Design Trust will work to implement *Five Borough Farm*'s policy recommendations. One initiative will be to develop a citywide process to make it easier for City agencies to identify land for urban agriculture. Working with Columbia University's Urban Design Lab, the Design Trust will help develop criteria to assess the suitability of city-owned land for urban agriculture. This process will enable agencies to review their land inventories systematically to meet citywide demand for farming and gardening, and to demonstrate that land is distributed equitably.

ACKNOWLEDGEMENTS

Five Borough Farm: Seeding the Future of Urban Agriculture in New York City would not have been possible without the generosity and expertise of many individuals. The Design Trust for Public Space and the Project Fellows would like to extend their sincere appreciation to all of the following for their assistance and thoughtful guidance.

Project Participants

Urban Farmers and Community Gardeners

John Ameroso, *Gericke Farms*
Elizabeth Bee Ayer, *BK Farmyards*
Bea Brown, *Drew Gardens*
Micheline Brown, *Warren/St. Mark's Community Garden*
Michelle Brown, *River Garden*
Harry Bubbins, *Friends of Brook Park*
Emmaline Carter, *211th Street Community Garden*
Anastasia Cole, *Brooklyn Grange*
Ray Figueroa, *Friends of Brook Park*
Ben Flanner, *Brooklyn Grange*
Sean-Michael Fleming, *The Secret Garden Farm & Nature Preserve*
Yonette Fleming, *Hattie Carthan Community Farm*
Wayne Fleshman, *Temple of David Community Garden*
Julie Forgione, *River Garden*
Regina Ginyard, *La Finca del Sur*
Joe Guimento, *River Garden*
Devanie Jackson, *Bed-Stuy Farm at the Brooklyn Rescue Mission*
Robert Jackson, *Bed-Stuy Farm at the Brooklyn Rescue Mission*
Kennon Kay, *Queens County Farm Museum*
Frieda Lim, *Slippery Slope Farm*
Ian Marvy, *Added Value*
Christian Mora, *EcoStation NY / Bushwick Farmers Market*
Will Morgan, *Little Green Acre*
Annie Moss, *La Finca Del Sur*
Stacey Murphy, *BK Farmyards / High School for Public Service*
Annie Novak, *Rooftop Farms*
Phyllis Odessey, *Randall's Island Park*
Steve Perry, *John Bowne High School*
Viraj Puri, *Gotham Greens*
Leah Retherford, *Queens County Farm Museum*
Kristin Schafenacker, *Added Value*
Philip Silva, *Prospect Heights Community Farm / Brooklyn Botanical Gardens*
Rocio Silverio, *La Finca Del Sur*
Jenna Spevack, *New York City College of Technology*
Abu Talib, *Taqwa Community Garden*
Vandra Thornburn, *Greene Acres*
David Vigil, *East New York Farms*

Karen Washington, *New York City Community Garden Coalition / La Familia Verde*
Jonathan Wilson, *Project EATS Active Citizen Project*
Sheila Yappow, *Essex Community Garden*

Nonprofit Organizations, Advocates, and Institutions

Kubi Ackerman, *Columbia University Urban Design Lab*
Holley Atkinson, *Slow Food NYC*
Jessica Bartolini, *Chhaya Community Development Corporation*
Andrew Bowman, *Doris Duke Charitable Foundation*
Daniel Bowman Simon, *TheWhoFarm (The White House Organic Farm Project)*
Wendy Brawer, *Green Map System*
Ursula Chanse, *Bronx Green-Up, New York Botanical Garden*
Skye Cornell, *Wholesome Wave*
Marissa DeDominicis, *Earth Matter*
Derek Denckla, *Farm City Fund*
Tanya Fields, *The BLK Project*
Rosalind Francis, *Harlem Seeds*
Taireina Gilbert, *Loisaida Inc.*
Christina Giorgio, *Urban Transformers*
Mara Gittleman, *Farming Concrete*
Denise Hoffman-Brandt, *City College of New York Graduate Landscape Architecture Program*
Hugh Hogan, *North Star Fund*
Michelle Hughes, *New Farmer Development Project, GrowNYC*
Debra Italiano, *Garden Labs NY / Garden Direct*
Nadia Johnson, *Just Food*
Kelli Jordan, *Farming Concrete*
Claudia Joseph, *Old Stone House*
Terry Kaelber, *Healthy Communities Through Healthy Food*
Erica Keberle, *GrowNYC*
Aley Kent, *Heifer International*
Susannah Laskaris, *landscape designer & garden services*
Lenny Librizzi, *GrowNYC*
Adam Liebowitz, *THE POINT Community Development Corporation*
Chaquita McCullough, *Harlem Seeds*
Sandra McLean, *Slow Food NYC*
Elizabeth Miller, *Doris Duke Charitable Foundation*
Joanne Morse, *Trust for Public Land*
Pamela Nathenson, *Fund for Public Health*
Jessica Nizar, *Fifth Avenue Committee*
George Pisegna, *Horticultural Society of New York*
Anne Pope, *Sustainable Flatbush*
Vicki Sando, *P.S. 41*
Marianna Schaffer, *David Rockefeller Fund*
Andy Stone, *Trust for Public Land*
Owen Taylor, *Just Food*
Sophia Vincent Guy, *Garden Direct*
Ashley Wilhite, *New York Restoration Project*
Toya Williford, *Brooklyn Community Foundation*

Rebecca Wolf, *Queens Botanical Garden*
Christine Yu, *NYC Food and Fitness Partnership*
Angelo Zaharatos, *Garden Direct*
Kolu Zigbie, *Jesse Smith Noyes Foundation*

Government Officials

Sabrina Baronberg, *Department of Health & Human Services*
Adrian Benepe, *Department of Parks & Recreation*
Rob Bennaton, *New York City Housing Authority*
David Bragdon, *Mayor's Office of Long Term Planning & Sustainability*
Randy Fong, *Department of Citywide Administrative Services*
Evan George, *Department of Parks & Recreation*
Cristina Grace, *New York State Department of Agriculture and Markets*
Liam Kavanagh, *Department of Parks & Recreation*
Kim Kessler, *Mayor's Office of Operations*
Margarita Lopez, *New York City Housing Authority*
Caroline Mees, *GreenThumb*
Whitney Reuling, *Garden to School Cafe*
Edie Stone, *GreenThumb*
Alison Tocci, *City Parks Foundation*
Angela Tolosa, *NYC Department of Corrections*
Lee Trotman, *New York City Housing Authority*
Elizabeth Tuckermathy, *United States Department of Agriculture*

Design Trust Team

Nevin Cohen, *Policy Fellow*
Agnieszka Gasparska, *Graphic Design Fellow*
Elliott Maltby, *Design Fellow*
Ian Marvy, *Added Value*
Gita Nandan, *Design Fellow*
Kristin Reynolds, *Research Collaborator*
Rupal Sanghvi, *Metrics Fellow*
Rob Stephenson, *Photo Urbanism Fellow*

Special Thanks to Former Design Trust Staff

Deborah Marton, *Executive Director*
Stephanie Elson, *Director of Programs*
Leslie McBeth, *Development Associate*

Project Advisors

Five Borough Farm Advisory Committee

John Ameroso, *Cornell Cooperative Extension*
Jacquie Berger, *Just Food*
Ray Figueroa, *Friends of Brook Park*
Yonette Fleming, *Hattie Carthan Community Garden*
Anthony Giancatarino, *Center for Social Inclusion*
Regina Ginyard, *La Finca del Sur*
Robert Jackson, *Bed-Stuy Farm at the Brooklyn Resuce Mission*
Aley Kent, *Heifer International*
Ian Marvy, *Added Value*
Annie Novak, *Rooftop Farms*
Steve Perry, *John Bowne High School*
Edie Stone, *GreenThumb*
David Vigil, *East New York Farms*
Karen Washington, *New York City Community Garden Coalition / La Familia Verde*

Policy Working Group

Sarah Brannen, *Center for an Urban Future*
Robert Jackson, *Bed-Stuy Farm at Brooklyn Rescue Mission*
Holly Leicht, *New Yorkers For Parks*
Edie Stone, *GreenThumb, Department of Parks & Recreation*
Karen Washington, *New York City Community Garden Coalition / La Familia Verde*

Metrics Working Group

Ray Figueroa, *Friends of Brook Park*
Nualla Gallagher, *Cypress Hills Development Corporation*
Mara Gittleman, *Farming Concrete*
Ian Marvy, *Added Value*
Stacey Murphy, *BK Farmyards*
David Vigil, *East New York Farms*

CONTRIBUTORS

Design Trust Fellows

Nevin Cohen

Policy Fellow

Nevin Cohen is an Assistant Professor of Environmental Studies at The New School, and Co-Chair of the Tishman Environment and Design Center, the university's interdisciplinary environmental research and education center. Dr. Cohen's research and professional work focuses on urban food policy in the urban and peri-urban landscape. He holds a Ph.D. in Urban Planning from Rutgers University, a Masters in City and Regional Planning from UC Berkeley, and a BA from Cornell.

Agnieszka Gasparska

Graphic Design Fellow

Agnieszka Gasparska is the creative director and founder of New York City-based design firm, Kiss Me I'm Polish, where she leads multidisciplinary teams in developing strategic and creative vision for websites, interactive applications, identity systems and print solutions. Before establishing her own firm, Ms. Gasparska spent five years as a senior art director at New York City's premier interactive agency, Funny Garbage. Her clients have included GOOD, The Museum of Modern Art, National Geographic and WW Norton. Learn more at http://kissmeimpolish.com/.

Rupal Sanghvi

Metrics Fellow

Rupal Sanghvi specializes in program evaluation, which means asking the question "does this work?" and then working with implementing programs to define meaningful metrics to address the question. In 2010, she founded HealthxDesign, which explores and communicates the role of design — including the built environment – for optimizing health outcomes. Ms. Sanghvi is also a Principal Investigator at the Public Health Institute and a consultant for the World Health Organization. Learn more at http://healthxdesign.org/.

Rob Stephenson

Photo Urbanism Fellow

Rob Stephenson is a fine art photographer based in Brooklyn, New York. His work focuses on the confluence of nature and the built environment. He has worked with award-winning New York photographer Joel Meyerowitz on several book projects, including 2009's "Legacy: The Preservation of Wilderness in New York City Parks." His photography has been exhibited at various venues including the Brooklyn Museum of Art, Jen Bekman Gallery, Lehman College Art Gallery and the Museum of the City of New York. Learn more at http://www.robstephenson.com/.

thread collective

Design Fellow

thread collective is an award-winning multi-disciplinary design studio in Brooklyn, New York founded by three graduates of U.C. Berkeley: Elliot Maltby, Mark Mancuso, and Gita Nandan. The collective explores the seams between building, art and landscape, stitching the diverse elements of the built environment to their ecological and social context through innovative design and research. The studio has served as a platform for collaboration with a diverse range of designers, artists, scientists, and policy makers. Learn more at http://www.threadcollective.com/.

Information Graphics and Book Design

Agnieszka Gasparska, Kiss Me I'm Polish

Photography

Rob Stephenson

Additional Contributors

Melissa Alexander, *Design Trust program associate*
Brian Brush, *mapping*
Chiara Di Palma, *Design Trust intern*
Rachel Dannefer, *interview coding*
Danielle Dowler, *mapping*
Michael Fox, *Design Trust intern*
Jennifer Gardner, *Design Trust intern*
Danielle Honey, *workshop coordinator*
Margaret Seiler, *copyeditor*
Do Mi Stauber, *indexer*

GLOSSARY

Administrative Code

A detailed collection of laws, codes, and ordinances that governs the activities of the City. The Administrative Code of the City of New York is organized into 30 titles according to subject matter.
http://72.0.151.116/nyc/AdCode/entered.htm

Aquaponics

A method of farming that combines aquaculture (the production of aquatic animals) and hydroponics (the cultivation of plants in water). Aquaponics uses plant life as a natural filtration system for a shared water supply that is continually recirculated for use by the aquatic animals.

City Charter

A document that establishes a city's system of governance. In New York City, the City Charter instituted a central system of government, with a Mayor-Council structure.

Combined Sewage Overflow (CSO)

New York City's sewer system combines rainwater and sewage. During even small storms, the system overflows, causing untreated sewage to flow into the city's waterways.

Commercial Farm

There are a handful of for-profit farms in the city, including three rooftop farms in the industrial areas of Long Island City and Greenpoint: Eagle Street Rooftop Farm, Brooklyn Grange, and Gotham Greens. In general, commercial farmers try to maximize crop performance in order to achieve profitability, in contrast with community farmers and gardeners. However, some commercial farmers share many of the goals of the broader urban agriculture community, such as promoting healthy eating and contributing to ecological benefits. Like community gardens and farms, commercial farms also rely heavily on volunteer labor and donated materials.

Community Board

Local representative bodies in New York City responsible for shaping municipal priorities in land use, advising on the services provided by municipal agencies, recommending agency budget expenditures, and evaluating the environmental impacts of proposed projects. Each of the City's 59 Community Boards consists of 50 unsalaried members.

Community Farms

For the purpose of this project, community farms are defined as growing spaces operated by a nonprofit organization (rather than by a group of volunteers, as is typically the case with community gardens) that includes in its mission engagement with the surrounding community. Typically, community farms feature communal growing spaces managed by staff and volunteers, rather than individual plots.

Growing food is an integral part of a community farm's operation. However, unlike commercial farms, community farms focus on community development and social programs such as school education programs and youth leadership training, rather than on maximizing food production and profitability.

Community Gardens

Gardens typically are run by a core group of volunteers or "members," usually drawn from the surrounding neighborhood, who maintain individual plots or communal growing spaces. There are approximately 490 community gardens citywide, covering just under 100 acres in total area. Most of these gardens are located on City-owned land and were formerly vacant lots, which proliferated in many neighborhoods of the city in the 1970s and 1980s. Gardens on city-owned land are required to host open hours for the general public.

Most of these gardens provide space for many different activities, including growing vegetables, growing flowers, or simply gathering for socializing or passive recreation. Roughly 80% of these gardens produce some food.

Compost

Compost is the product of the decomposition of organic matter. When mixed with soil it provides nutrients and improves soil structure. Finished compost can be produced in less than one year through properly managed decomposition of green plant materials and kitchen scraps, combined with dry organic material such as wood chips and paper.

Community-Supported Agriculture (CSA)

A method of food production and delivery in which members pre-pay for a share of a particular farm's produce. Usually organized around a growing season, a CSA share might include fruits, vegetables, herbs, flowers, eggs, meat, poultry, dairy products and/or honey.

Community Development Block Grants (CDBG)

A federal program of the the Department of Housing and Urban Development (HUD) that funds local community development activities, including the GreenThumb program.
http://portal.hud.gov/hudportal/HUD?src=/program_offices/comm_planning/communitydevelopment/programs

Cornell Cooperative Extension

A Cornell University program that provides educational programs and resources to New York state residents, businesses, and organizations, in every county of the state. Program areas include: agriculture, public health, natural resources, land-use, workforce and youth development.
http://www.cce.cornell.edu/Pages/Default.aspx

Environmental Protection Agency (EPA)

A Federal agency responsible for protecting the health of the nation's human population and the environment. The EPA develops and enforces environmental regulations based on laws passed by Congress.

Farm Bill

A comprehensive bill that addresses affairs of the U.S. Department of Agriculture. This is the primary policy tool of the Federal government with regard to food production in the United States.

Farming Concrete

A community-based research project to measure how much food is grown in New York City's community gardens and school gardens.
http://farmingconcrete.org/

Food Access Programs

Programs such as the Supplemental Nutrition Assistance Program (SNAP) or Health Bucks that strive to make healthy, affordable food available to all.

Food Hubs

Facilities that provide equipment and space to refrigerate, process, and distribute food.

Food Policy Coordinator

Created by the Bloomberg Administration in 2007, this position is responsible for improving food quality and access to healthy food throughout New York City. The Food Policy Coordinator, who reports to the Director of Health Services within the Office of the Deputy Mayor for Health and Human Services, primarily focuses on the needs of low-income citizens with limited access to healthy food.

FoodNYC

FoodNYC: A Blueprint for a Sustainable Food System is a food policy white paper issued in February 2010 by Manhattan Borough President Scott M. Stringer. *FoodNYC* prioritizes access to healthy food and local food production as strategies in service of job creation and improving both human and environmental health.
http://www.mbpo.org/release_details.asp?id=1496

FoodWorks

A comprehensive, citywide food policy platform issued by City Council Speaker Christine C. Quinn in November 2010. *FoodWorks* offers a detailed analysis of New York City's food system, from production to waste management, including a section with policy proposals devoted to food production.
http://council.nyc.gov/html/action_center/food.shtml

Green Carts

A program of the Department of Health and Mental Hygiene supporting 1,000 licensed mobile food vendors to improve retail access to fruits and vegetables in specific underserved communities.
http://www.nyc.gov/html/doh/html/cdp/cdp_pan_green_carts.shtml

Green Infrastructure Plan

Released by the New York City Department of Environmental Protection in 2010, this stormwater management plan outlines low-tech, landscape design techniques to slow and absorb stormwater runoff to reduce combined sewage overflows into the city's waterways.
http://www.nyc.gov/html/dep/html/stormwater/nyc_green_infrastructure_plan.shtml

GreenThumb

A program within the New York City Department of Parks & Recreation that provides support to urban farms and gardens in New York City. GreenThumb licenses community gardens located on city property, offers technical assistance, supplies a number of different material resources, and provides labor to help clear vacant lots for new gardens and for garden improvement projects.
http://www.greenthumbnyc.org/

GrowNYC

An independent nonprofit organization that operates farmers markets, administers environmental educational programs, funds garden improvement projects, and creates community gardens in New York City.
http://www.grownyc.org/about

Health Bucks

Developed and distributed by the New York City Department of Health and Mental Hygiene, Health Bucks are coupons worth $2 that can be redeemed at farmers markets by recipients of the Federal Supplemental Nutrition Assistance Program (SNAP) to purchase fruits and vegetables.
http://www.nyc.gov/html/doh/html/cdp/cdp_pan_health_bucks.shtml#hb

Horticultural Society of New York (HSNY)

A nonprofit organization focused on horticultural education and fostering a community of urban gardeners in New York City.
http://thehort.org/aboutus_missionhistory.php

Horticultural therapy

The use of gardening activities, facilitated by a trained therapist, for the purpose of achieving specific treatment goals. Horticultural therapy has been used to treat patients recovering from stroke and heart disease, blind and partially sighted people, those in the early stages of dementia, and people with physical and learning disabilities.
http://www.ahta.org/content.cfm?id=history

Indicators

Indicators are signs of progress and change that result from an activity, project, or program. Indicators can be compared to road signs which tell travellers if they are on the right road, how far they have traveled, and how much distance is left to reach a destination.

In-kind

Providing services or donating goods in lieu of a financial contribution.

Institutional Farm

These include farms and gardens at hospitals, churches, prisons, supportive housing developments, and other institutions whose primary mission is not food production, but which have goals that urban agriculture supports.

Land Trust

A nonprofit organization that conserves land. Three land trusts in New York City (Manhattan, Bronx, and Brooklyn Queens) have title to more than sixty community gardens, and work to ensure long-term stewardship and community-based control of these spaces.

Mayor's Management Report

An annual report from the Mayor to the public and the City Council on the performance of 44 municipal agencies and organizations in delivering services.
http://www.nyc.gov/html/ops/html/data/mmr.shtml

Mayor's Office of Environmental Coordination

Assists City agencies in carrying out their environmental review responsibilities.
http://www.nyc.gov/html/oec/html/home/home.shtml

New York City Department of Buildings (DOB)

A government agency that enforces the Building Code, Zoning Resolution, and other applicable laws through a process of construction permitting and site visits. DOB issues building permits for installations such as green roofs.
www.nyc.gov/html/dob

New York City Department of Citywide Administrative Services (DCAS)

The Department of Citywide Administrative Services is responsible for all city-owned properties. Local Law 48 of 2010 requires DCAS to publish a list of all city-owned properties that are suitable for urban agriculture.
www.nyc.gov/html/dcas

New York City Department of Correction (DOC)

A government agency that is responsible for the care, custody, and control of persons either accused of crimes, or convicted and sentenced to one year or less of jail time. DOC operates an institutional farm on Riker's Island.
www.nyc.gov/html/doc

New York City Department of Education (DOE)

A government agency that provides education to primary and secondary students in over 1,400 schools across the five boroughs. DOE's SchoolFood Division helps administer Garden to School Café, which uses produce grown in school gardens in cafeteria meals.
http://schools.nyc.gov

New York City Department of Environmental Protection (DEP)

A government agency that manages and conserves the city's water supply, providing over one billion gallons of drinking water each day. DEP is also the local entity in charge of enforcement of Federal Clean Water Act rules and regulations. DEP's Green Infrastructure program promotes strategies to absorb rainwater, including transforming paved vacant lots into urban farms and gardens.
www.nyc.gov/html/dep

New York City Department of Health and Mental Hygiene (DOH)

A government agency that provides comprehensive public health services to protect and promote the health of all New Yorkers. DOH develops and distributes Health Bucks.
www.nyc.gov/html/doh

Department of Housing and Urban Development (HUD)

A federal agency that provides quality affordable housing options, and administers Community Development Block Grants.
http://en.wikipedia.org/wiki/HUD

New York City Department of Housing Preservation and Development (HPD)

A government agency that utilizes various tools of preservation, development, and enforcement to improve the availability, affordability, and quality of New York City housing options.
www.nyc.gov/html/hpd

New York City Department of Parks & Recreation (DPR or "Parks")

A government agency that maintains and operates city-owned open space amenities and recreational facilities, and manages the GreenThumb program. More than half of the city's 490 community gardens are under Parks' jurisdiction.
www.nycgovparks.org

New York City Department of Sanitation (DSNY or "Sanitation")

A government agency that manages solid waste. Sanitation's responsibilities include producing compost and funding community-based compost education and outreach programs.
www.nyc.gov/html/dsny

New York City Department of Transportation (DOT)

A government agency that is responsible for the maintenance of streets, bridges, and traffic signals, street signs, parking meters, and streetlights in the city.
www.nyc.gov/html/dot

New York City Economic Development Corporation (NYCEDC)

A local development corporation that encourages citywide economic growth.
www.nycedc.com

New York City Housing Authority (NYCHA)

The public housing authority that provides affordable housing to low and moderate income City residents in housing developments and apartment buildings borough-wide. NYCHA's Garden and Greening Program helps public housing residents maintain some 600 gardens citywide.
www.nyc.gov/html/nycha

New York State Department of Agriculture and Markets

A government agency that promotes agricultural industry and environmental stewardship statewide. The department supports programs to assist community gardens and to enable low-income New Yorkers to purchase food from farmers markets.
http://www.agriculture.ny.gov/

New York State Department of Environmental Conservation (DEC)

A government agency that conserves, improves, and protects New York's natural resources and environment through prevention, abatement and control of water, land, and air pollution.
http://www.dec.ny.gov/

Office of Management and Budget

A City agency responsible for assisting the Mayor in developing and implementing the City's budget, and for advising the Mayor on policy affecting the City's fiscal stability and the effectiveness of City services.
http://www.nyc.gov/html/omb/html/home/home.shtml

Phytoremediation

The use of plants to mitigate polluted contaminants in soil, water or air.

PlaNYC

New York City's citywide sustainability plan, released by Mayor Bloomberg in 2007, and updated in 2011. The update includes goals and specific objectives to support and expand urban agriculture.
http://www.nyc.gov/html/planyc2030/html/home/home.shtml

Raised bed

An open wood frame filled with soil that allows farmers and gardeners to grow in healthy soil above the ground.

Soil remediation

Various strategies used in the purification and/or revitalization of soil.

Stormwater

Water that originates during rain, snow, or other precipitation events. Stormwater is easily polluted as it flows over the ground.

Stormwater runoff

Stormwater that is not absorbed into the ground, and that can flow into sewer systems or waterways. Stormwater can pick up debris, chemicals, and sediment, which contaminate other water sources if left untreated.

Raised bed

An open wood frame filled with soil that allows farmers and gardeners to grow in healthy soil above the ground.

Supplemental Nutrition Assistance Program (SNAP)

A federal program of the U.S. Department of Agriculture that provides financial assistance to low-income individuals and families for purchasing food. SNAP was formerly known as the Food Stamp Program.
http://www.fns.usda.gov/snap/

United States Department of Agriculture (USDA)

A federal government agency responsible for developing and executing policy on agriculture, farming, and food at the national level.
http://www.usda.gov/wps/portal/usda/usdahome

ULURP

The Uniform Land Use Review Procedure (ULURP) ensures that New Yorkers can review proposed land use changes, from the sale of City-owned property to the designation of urban renewal areas, before land is conveyed or designated. ULURP mandates that the affected community board and the Borough President consider and vote on land use changes before they can be approved by the City Planning Commission and—for some actions—by the City Council.
http://www.nyc.gov/html/dcp/html/luproc/ulpro.shtml